ALSO BY DAVID WILSON

Body and Antibody: A Report on the New Immunology (1972)

This is a Borzoi Book, published in New York by Alfred A. Knopf

THE NEW ARCHAEOLOGY

The New ARCHAEOLOGY

David Wilson

ALFRED A. KNOPF New York 1975

THIS IS A BORZOI BOOK
PUBLISHED BY ALFRED A. KNOPF, INC.

 Published in the United States by Alfred A. Knopf, Inc., New York. Distributed by Random House, Inc., New York. Originally published in Great Britain as *Atoms of Time Past* by Allen Lane, a Division of Penguin Books Ltd., London.

ISBN: 0-394-47936-X
Library of Congress Catalog Card Number: 74-21321

Manufactured in the United States of America

First American Edition

Contents

THE NEW ARCHAEOLOGY

1

Science and Archaeology

Archaeology began as part of science. Just over one hundred years ago archaeology achieved its first great triumph by providing the evidence that convinced the majority of men of the validity of the most revolutionary theory proposed by science. The theory was the theory of evolution, put forward by Charles Darwin in his book *The Origin of Species*. The theory of evolution carried the implication that, in physical terms at least, man was part of the natural world he saw about him; that man had emerged from the animals in the same way that the different species of animals had emerged from other ancestral species; that the process of emergence had taken millions of years to occur on a planet that was thousands of millions of years old. Accepting this theory meant, to the men of 1850, a complete change in man's concept of himself and a complete change in the perspective of the universe in which he lives.

The evidence that compelled most people to accept the shattering of man's traditional picture of himself was the demonstration that some men had lived at the same time as animals of types that did not exist in 1850. These animals were related to the animals of our modern age, yet they were clearly different; the only explanation was that they must be extinct types of animals, ancestral to the present forms. And therefore the men who lived alongside them must have lived many thousands of years before the beginning of history as it was known in 1850. The men who provided this evidence dug into the ground to get it. They were the first archaeologists, though they did not call themselves so. They regarded themselves as scientists or, more specifically, as geologists.

In the hundred years that archaeology has existed, however, it has moved far away from its origins in that culture that we call 'science', and it has become part of that other culture (is it a rival culture or a complementary one?) which we call 'the humanities', or 'the arts'. And now there is a violent swing back again towards science, so that many an archaeologist today will be offended if he is not accepted as a scientist, and a good number of undoubted scientists are devoting most of their energies to providing evidence in archaeological problems.

The major concern of this book is with the 'swing' back towards science that has taken place within archaeology. This movement has been caused, at least partly, by a whole series of discoveries that the techniques of other branches of science, the techniques of what we may call 'laboratory science', can be applied to archaeological material so that archaeology itself can provide much more knowledge about the past. In turn this vast increase in the 'productivity' of archaeology has led to changes in our views about what actually occurred during the long history of man which are almost as revolutionary as the discovery that man *had* a long history. With this have come changes in the archaeological profession and in the archaeologists themselves. And, most recently, there has come the movement called 'the new archaeology', which proposes that archaeology must not only use scientific techniques, but must also adopt the techniques of scientific method and scientific thinking throughout its operations.

In a sense, then, archaeology has come the full circle since its birth in the 1850s, when men, who simply considered themselves to be scientists, used archaeological techniques to provide evidence for a scientific theory which was of far wider import than the history of man alone. The evidence which showed that man had a very long history was archaeological not only in the sense that it was dug out of the ground. The evidence was found in the ground by digging and excavation, but the important features – the facts that made the findings into evidence about what occurred in the past – were the relative positions and depths of the various items that were found and the nature of the ground, earth and rock in which they were found.

In the 1850s this evidence was termed 'evidence for the existence of Antediluvian Man', which, strictly speaking, meant evidence for the existence of man in northern Europe before the date traditionally accepted for the Great Flood of the Bible. In fact the term 'Antediluvian Man' soon came to mean, broadly, man who had lived many thousands of years before. The fact that this evidence was discovered in northern Europe – in Britain and France – was probably important in obtaining rapid acceptance of this idea among the most influential men of the time. Northern Europe was, at that date, the intellectual centre of the world, and the evidence was available for inspection only a short distance from London, Paris and Berlin. There was no necessity to travel as far as the Galapagos Islands to see for oneself that Darwin was right.

It was in 1838 that Jacques Boucher de Perthes, a French customs official, working at Abbeville, first tried to persuade the larger world of scientists and antiquarians that the curiously-shaped pieces of flint which he had discovered in the gravel pits of the valley of the river Somme were tools made by men. And since these 'stone-axes' were found lying many feet below the contemporary surface of the earth, at the same levels as the bones of animals of types which were clearly extinct, these men must have lived at the same time as the extinct animals, many thousands of years before the date at which the world was generally supposed to have been created. It took Boucher de Perthes twenty-one more years of digging and collecting, of writing and publishing books and pamphlets, of confrontations and arguments, before his theories were accepted.

In France the crucial swing of opinion towards Boucher de Perthes' theories came after one of his avowed opponents, Dr Rigollet of Amiens, started independent excavations of his own in the gravel pits of St Acheul, in an attempt to disprove the assertions of man's great antiquity. He dug up so many flint tools and bones of extinct animals, however, that he was forced to change his ideas. In his final report, which he published in 1854, Dr Rigollet publicly proclaimed his conversion to Boucher de Perthes' theories. His intellectual honesty has been rewarded because an entire period of man's culture-history is now called Acheulian, and wherever archaeologists recover flint-tools of the type first found by Dr Rigollet they are given this name derived from the village gravel-pit.

In these same years the same controversy sprang up on the other side of the Channel, based on similar finds and claims in southern England. The controversy centred round a natural cave, called Kent's Cavern, near the little seaside town of Torquay in Devonshire. Two excavators, a Catholic priest, Father MacEnery and Mr Godwin Austen claimed to have found the bones of rhinoceros and cave bear together with flint tools that must have been made by men, sealed in under a layer of stalagmite. But the cave was well-known locally, so that they could not prove that man had not dug through the stalagmite at some date later than the lifetime of the animals and dropped or left his tools in the holes. The work was started again in 1846 by William Pengelly, the son of a Cornish fisherman, the very exemplar of the self-educated man of Victorian idealism, who ended his life as a Fellow of the Royal Society. He re-started the excavation of Kent's Cavern, primarily as a fieldwork project for the local Natural History Society he had been instrumental in founding. In a few years' work Pengelly convinced himself that the earlier excavators had been right and that the flint tools were truly contemporary with the bones of animals that were either extinct or that had not inhabited England for thousands of years. But he could not prove absolutely to others that there had been no intrusions through the layer of stalagmite in later times.

It was in 1858 that the entrance to a previously unknown cave was revealed by quarrying work at Windmill Hill above Brixham Harbour – a few miles away from Kent's Cavern on the other side of Torbay. The Geological Society and the Royal Society organized an official excavation. They called on Pengelly to do the work and set five of the most eminent geologists of the country to supervise and watch him. In due course Pengelly's workmen revealed a layer of stalagmite at the bottom of the cave. It was carefully cleared and then examined by Pengelly and the members of the supervisory committee. It was clearly unbroken. It did not take long after that to find remains in the shape of teeth and bones of mammoth and woolly rhinoceros, of cave bear and hyena and reindeer. The animal remains were not only in the layer of stalagmite but underneath it, too. And the remains of man, undeniable flint tools, were alongside and underneath the teeth and bones of animals extinct or of such species as were normally found in the tropics or in the Arctic, but never in Britain.

The work at Windmill Hill was finished in the early spring of 1859. Two members of the supervisory committee of (now convinced) geologists visited Boucher de Perthes and his diggings later that spring. The meetings of the Royal Society, the British Association for the Advancement of Science, and the Royal Society of Antiquaries of 1859 all accepted publicly that there was incontrovertible evidence for allowing man to be of great antiquity as a species. So great was this antiquity, so long had man existed, that all thinking about the human race had now to be put into a different perspective.

None of the learned societies that formally vindicated Pengelly and Boucher de Perthes was an archaeological organization. None of the men who supervised and checked Pengelly's work was an archaeologist – they were geologists and Pengelly himself turned next to geological work. The people who collected 'buried' objects in the 1850s, and sometimes even dug for them, were 'antiquarians'. Pengelly and Boucher de Perthes deserve the credit for being the first true, scientific, archaeologists, but their work, and that of the other scientific 'fossil hunters' of their time, simply opened up an enormous gap in man's account of himself. Man had apparently existed for many thousands of years – yet men only knew of the Romans and Greeks and Egyptians and Jews stretching back at most some five or six thousand years. What had been regarded as the whole of history leading up to the 1850s was now seen to be no more than the spiritual and intellectual history of our western culture. What had happened to all the rest of mankind for all the other many thousands of years?

The answers to these questions were just beginning to be glimpsed by the 'antiquarians' at the time. We can now see that antiquarians provided the other two main strands in the history of archaeology that intertwine with the strand of purely scientific investigation. Science had shown that the 'time element' in man's history was on a scale vastly larger than anyone had conceived. The 'antiquarians' were now starting to put a 'time element' into the miscellaneous collection of objects which northern Europe had inherited from the past but had made little attempt to understand. This ordering of antiquities into a time sequence to fill the gap opened up by Pengelly and Boucher de Perthes can be considered as the second main strand of archaeology.

It began in Scandinavia, and particularly in Denmark, where the peat bogs have the peculiar property of sometimes preserving articles or creatures virtually intact for thousands of years. (The discovery of the perfectly preserved body of the Iron Age individual called 'Tollund Man' in our own time is a reminder of this.) Many of the things found in Denmark over the years had been collected together in the Royal Cabinet of Antiquities – which was just what its name implies, a small room or large cupboard in the King's Palace in Copenhagen in which these curious objects were stored. After some years of pressure from various 'savants' and antiquarians it was decided, in 1819, to display these objects to the public in a Museum of Danish Antiquities, and the man appointed to organize the display was Christian Thomsen. He was a well-to-do middle-class young man who had become interested in antiquities when he helped to rescue a private collection of coins from the flames caused by Lord Nelson's bombardment of Copenhagen in 1807. Thomsen's primary employment was in his father's business, and he organized the Museum of Antiquities in an honorary capacity. When he eventually opened it, it was seen that the first display cabinet contained objects made of stone, the second objects made of bronze, and the third showed 'antiquities' made of iron.

At the time of the opening of the Museum, Thomsen had probably not fully committed himself to the idea that this division by material also involved a division in time. But over the next twenty years Thomsen slowly developed the idea until the full concept, which we now accept, took shape – and prehistory was divided into three main time-periods, the Stone Age, the Bronze Age and the Iron Age. It was not until 1836 that Thomsen was prepared to commit himself fully to this idea, but then it was announced to the world in the shape of a single chapter which he contributed to a book which is now usually known as the *Ledetraad*. The *Ledetraad* was the *Guide to Scandinavian Antiquities*, and Thomsen's chapter was officially no more than a description of the Copenhagen Museum, but this involved describing the Three-Age System and its justification. He went further and described how objects made of other materials, such as bone or pottery, could be ascribed to Stone or Bronze or Iron Age by distinguishing the forms of art used in the different periods.

The Three-Age System was not acceptable immediately to the intellectual world of the 1830s. There were other schools of thought, such as that led by Professor Nilsson of Lund University in Sweden, though he eventually supported Thomsen. Outside Scandinavia, as the *Ledetraad* was eventually translated into French, German and English, the controversy was even more violent. It was the work of Jens Jacob Asmussen Worsaae in the 1840s that showed that Thomsen's museum-based Three-Age System corresponded well with excavation in the field – in other words Stone Age objects would be found buried under Bronze Age objects, and had therefore been made and used before Bronze Age objects. Iron Age objects came uppermost and therefore most recently. Worsaae, who trained under Thomsen and followed him at the Danish Museum, is often described as the first full-time, paid archaeologist in history. But it has recently been discovered that Leiden University in Holland had the equivalent of a full-time professor of archaeology – a man named Reuvens – as early as 1819. Nevertheless, there is no doubt that the main impetus for establishing the Three-Age System, the first time-base for prehistory, came from Denmark in both museum work and Worsaae's field excavation. And as the *Ledetraad* spread so did Thomsen's ideas – they were taken up by powerful and influential scholars like Lord Avebury in Britain – though even when Thomsen died, in 1865, there were still many scholars who rejected his theory.

It was in exactly these same decades of the nineteenth century – the 1840s and 1850s – that the third strand which formed archaeology was being spun. Again it was primarily an 'antiquarian' strand. But the antiquarians in this case were not dull museum curators, but lively and adventurous, romantic and quarrelsome, travellers and explorers. It can be claimed that Napoleon began it all when he invaded Egypt, and introduced Egyptian and Middle Eastern ideas and motifs to the imagination and fashion of western Europe. By the 1840s and 1850s Henry Austen Layard, on behalf of the British Museum was fighting his French rivals in Mesopotamia for the sculptured bulls and lions of Nineveh as fiercely as Wellington had fought Napoleon's marshals in Spain and Portugal. From underneath the enormous sandy mounds of the country we now call Iraq the remains of the Babylonian and Assyrian civilizations were being dragged out to fill the museum galleries of London and the Louvre. The first great intellectual impact of the ruins of the Mesopotamian palaces was felt in the early 1850s and Layard's most important book about his excavations was published in 1855.

So the situation arose that at exactly the time when science, aided by archaeology, was showing that man's history stretched back many thousands of years and must therefore contain far more than the development of the classical civilizations of Egypt, Greece and Rome, the activities of the traveller-antiquaries in the Middle East were showing that enormous civilizations and empires – barely mentioned in the Bible, dismissed by the Greeks as 'barbarians' – had existed alongside what had, up to that time, been considered as the mainstream of human history. Most educated men in western Europe and America had been brought up on a mixture of Roman, Greek and biblical studies, so the extensions of knowledge to include the Babylonian, Assyrian and

Egyptian civilizations came as no great difficulty. The marvellous and colossal carvings of Mesopotamia were attractive in themselves and far more dramatic and stimulating to the imagination than the dull flints and battered and corroded metalwork of the Three-Age System of the Scandinavian museum curators. So the third strand of archaeology dominated the other two. The unearthing of Middle Eastern antiquities rapidly developed into scholarship, and the new discipline of archaeology was dominated by the scholarship of unearthing the history of the Middle Eastern civilizations. Knowledge of these civilizations, which had preceded the cultures of Greece and Rome, was manifestly an extension of knowledge of the classical civilizations. Thus archaeology became one of the 'humanities'. It was certainly treated as such in almost all the universities of western Europe and America, not only in the sense of sharing the cultural values of classical and humane studies, but also in the formal and administrative sense – the department of archaeology normally was part of the faculty of arts or was grouped under the 'classical studies', or whatever nomenclature might be chosen by any particular university.

In the century that has elapsed between the birth of archaeology and our own time, diggings and excavations all over the world have moved us into the position where we now know far more about our distant ancestors and our origins than did the many generations who lived before us and who were, in terms of time, much closer to those distant ancestors and their cultures and civilizations than we are. All groups, tribes or nations give some account of their past and origins but it is a strange fact that it is only in recent times that it has seemed in any way important to us to get our history reasonably correct and free from myth. (There have, of course, been notable exceptions to this, Herodotus, Thucydides, Bede, but these have been isolated individuals.)

The Greeks certainly knew something about their own origins. Homer, whoever he or they may have been, clearly understood that he was writing in an Iron Age about the exploits of warriors whose armour and accoutrements consisted almost entirely of bronze. But of a Stone Age preceding the Bronze there is no mention and other ancient Greek authors have myths of ages of gold and silver at the start of history. It is claimed by Jacquetta Hawkes that the first recorded piece of archaeological excavation and deduction is to be found in the pages of Thucydides, who wrote,

The Islanders were Carians and Phoenicians by whom most of the islands were colonised, as was proved by the following fact. During the purification of Delos by Athens in this war all the graves in the island were taken up and it was found that half their inmates were Carians. They were identified by the fashion of the arms buried with them and by the method of interment which the Carians still follow.

But the story of the discovery of an ancient bronze tablet in a tomb at Haliartus which was sent off to the Egyptian priests because no one could read it, a story told by Plutarch, combines with much other evidence to show that the Greeks of the classical period had lost all memory and knowledge of the Cretan language, writing and civilization which contributed so much to their own culture only a few hundred years before the glorious, high peaks of Athenian civilization.

It is fifteen hundred years since the Romans left Britain, but we know more about Roman Britain now than our ancestors did eight hundred years ago. We have quite a good idea about how much the men of 1200 A.D. knew about the Romans in Britain. Geoffrey of Monmouth's chronicle *History of the Kings of Britain* had just been written, and it was clearly known that the Romans had ruled the country at some time in the past. The chronicle provides the first crystallization into writing of the misty Celtic legends of 'King Arthur', who was linked with the Romans and who appears to be their successor in battling against the invading Saxons. Yet, despite this knowledge, there was no appreciation in 1200 that the Romans had left any physical remains of their presence in Britain. 'Bad' King John, who ruled in the years immediately after 1200, was an active and energetic monarch, and one of the few who ever visited the north of England for any purpose except fighting the Scots. When he saw the remains of Hadrian's Wall, and the ruins of the Roman camps and fortresses that protected the northern boundary of the Roman Empire against the Picts and Scots, he believed that these mighty remains had been constructed by an unknown race of giants and had been destroyed by earthquakes. It was three hundred years later, in the reign of Henry VIII, before anything approaching what we would consider a 'true' history of England started to appear. And three hundred years after that, in the first half of the nineteenth century, just before archaeology was born, the antiquarians were still compelled to label anything that was pre-Roman simply as 'Ancient Briton' or 'Druid'. This ignorance was exactly paralleled in France, where anything ancient and non-Roman was called 'Gaulish', while all that the Germans knew of the early history of their country was what Tacitus could tell them about the attempts of the Romans to hold the Rhine.

Yet today, thanks largely to the archaeologists, we have a detailed knowledge of Hadrian's Wall and an often intimate picture of the lives of the men who patrolled it. On any fine day in summer hundreds of visitors wander among the ruins and ramparts, examining the bathing and sanitary arrangements in cavalry barracks or wondering how the Spanish and Syrian auxiliaries stood up to the winter winds that whistle across the bare uplands.

It is difficult for any of us, now, to see how men of the first half of the nineteenth century, men as intelligent as we are, men as well educated by the standards of their times as any of us, can possibly have been satisfied with the enormous gaps in their knowledge of the past. But the men who started archaeology were not, as a first objective, trying to fill these gaps. The three groups of men who founded archaeology had other objectives. The antiquarians of western Europe, mostly gentlemen who dug up the barrows and earthworks on their country estates in Denmark, Germany, France and England, were as much in search of treasure as anything; so too were the antiquarians of the Orient, the men who collected 'antiquities' in Mesopotamia or Egypt to send back to museums at home and who thus unearthed what anybody could see were the remains of civilizations quite unknown or known only through a passing mention in the Bible; and finally there were the scientists, primarily geologists, who dug to solve one problem only, at first, the problem of the antiquity of man in the general scheme of nature.

The situation is well summed up by Jacquetta Hawkes:

> During the last phase of the struggle for 'Antediluvian Man', there was a simultaneous development in the non-controversial exploration of past civilizations. Indeed, although both to some extent relied on excavation and might be seen by the broadminded to have a single purpose – the recovery of man's past – these two wings of archaeology, the scientific investigation of origins and the scholarly and humanistic appreciation of advanced cultures, appeared at this time to be almost in opposition. The centre of interest for the humanists had shifted from the Mediterranean of the classical world, to the great valley of the Two Rivers – the Mesopotamia which we now have to call Iraq; also, though less strongly, to the Nile Valley, which the cracking of the secret of hieroglyphs had now brought within the range of history. This change meant that exploration was focused on regions involved in Hebrew history – the 'lands of the Bible'. There is no doubt that as biblical cities, and texts related to Bible accounts, were found and identified, many naive or worried people saw them as opposing the unwelcome advances of the evolutionists. For those who did not want to distinguish between the religious myths of South West Asia and the history of the Hebrew people, the discoveries in Mesopotamia and Egypt quite simply 'proved the Bible to be true'. So the work of the explorers was followed with a supercharged interest and concern.[1]

In the Americas, of course, the white European invaders of the sixteenth century had actually met the Stone Age cultures of the North American Indians and the Aztec and Inca civilizations when they were still alive and existent. Aztec and Inca can, broadly speaking, be compared with the Mesopotamian and Egyptian Bronze Age civilizations except that the American versions had not such an advanced technology as the south-west Asians. The pattern of development of 'knowledge' of the history of the pre-European cultures of the Americas was remarkably similar to the development in Europe, except that there was a time lag of thirty years or more.

First, there were the 'antiquarians' of North America, and their work was well under way by the middle of the nineteenth century – with great publications like *Ancient Monuments of the Mississippi Valley* by Squiers and Davis in 1848 and *Archaeology of the U.S.* by Samuel Haven in 1855. But these 'antiquities' of the North American continent did not throw any light on the problems of the antiquity of man; they merely revealed some details of the unwritten history, the 'prehistory', of 'the Indians', rather in parallel to the European discoveries that the pre-Roman inhabitants of north-west Europe could not simply be lumped together as Druids and Gauls.

The exploration of the high civilizations of Central and South America was started, much as the exploration of Middle Eastern civilizations had been, by wealthy, aristocratic and often eccentric Europeans who came primarily as 'travellers'. Charles IV of Spain sent Dupaix and Castenada to report on the Mexican ruins in 1805; but their reports were lost in the disintegration of the Spanish monarchy under the impact of Napoleon. Some of their work was rediscovered thirty years later by Viscount Kingsborough, the heir to an Irish nobleman. He was convinced that the ten lost tribes of Israel had founded the mysterious jungle-clad cities of Central America, and he devoted his fortune and life to publishing in magnificent form all the ancient Mexican manuscripts and the up-to-date drawings and reports of Dupaix that he could find. Kingsborough gave up his seat in Parliament to pursue his work; he ruined himself financially and died of typhus in a Dublin debtor's prison at the age of forty-two, before he could inherit his father's peerage and vast fortune. He was followed by a man who lived to be 102, the German-French Comte de Waldeck.

Waldeck favoured the Egyptians, rather than the ten tribes of Israel, as the founders of Central American civilization, but that was probably because he had accompanied Napoleon on his conquest of Egypt. He visited Guatemala and examined many ruins and published lavishly. William Prescott's two great historical studies on the conquests of Mexico and Peru again drew attention to the extraordinary ruins to be found there, and while he was writing them the first real exploration of the sites was started by the adventurous John Stephens of New Jersey, accompanied by his English friend and illustrator, Frederick Catherwood. Stephens managed to get himself appointed Special Confidential Agent in Central America by the Washington government, and all his archaeological exploration in Central America was covered by this extraordinary title. He played much the same role in Central American exploration as did Henry Layard in Mesopotamia; both having more or less official government appointments, and both having the characteristics and literary skill to arouse public attention by their exploits and thus bringing their important discoveries to wide notice. Stephens' travels were between 1839 and 1842 yet it was not until the 1880s that really scientific investigation of the Central American civilizations got under way, and the first major expedition was not launched until 1888 when the Peabody Museum at Harvard pioneered the field, a full thirty years after serious European exploration of Middle Eastern sites had started. Serious investigation of the antiquity of man on the American continents can hardly be said to have gained much impetus until the discovery of the Folsom site in New Mexico, by the Negro cowboy, McJunkin, in 1925.

This was almost exactly at the time when Pekin Man was discovered in the limestone caves of Chou-kou-tien, largely as a result of the enthusiasm of the Canadian, Dr Davidson Black, supported by very large grants from the Rockefeller Foundation. The discovery of Pekin Man belongs in the story of the search for the origin of *homo sapiens*, but it was not until after he had been discovered that the Swedish geologist and explorer, Gunnar Andersson, who had played a part in opening up the Chou-kou-tien discoveries, started the first explorations of Neolithic and Bronze Age sites, the prehistory of China, in the early 1930s. The fact that India, too, had a prehistory of much distinction had emerged only a few years before, in the middle 1920s, when Sir John Marshall, excavating at Mohenjo-Daro and Harappa, in the

Punjab, revealed civilizations which had been even more completely obliterated and forgotten than the early civilizations of Mesopotamia.

The first clue to the existence of primitive humanoid fossils in China had come from the recognition of a fossilized human molar tooth in a Chinese 'chemist's shop', as part of a display of 'dragon's teeth' for medicinal purposes. The clue to Mohenjo-Daro came from the discovery of bricks in an apparently natural small hill which was being dug to provide ballast for railway construction.

Spotting these two odd clues, which both led to discoveries of the greatest importance, was probably the last major contribution of the 'antiquarians' to modern archaeology. But their tradition of interest in relics of the past lives on among the vast number of amateur 'helpers' and 'diggers' who can invariably be counted upon to help the professional archaeologist when it is necessary to organize a rescue dig, as the bulldozers of reconstruction threaten some ancient site in Britain or North America. And the 'antiquarian' tradition had also contributed, before the twentieth century began, one more essential element to the fabric of archaeology. The first 'scientific' excavators were interested in finding evidence to prove a point – the antiquity of man; the first humanistic excavators were interested in obtaining 'works of art' and 'antiquities' for museums. It was left to the greatest of the antiquarian amateurs to show how the actual dig should be conducted and recorded in a truly 'scientific' manner, aware that things he found and the places in which he found them might mean more to his successors, re-reading his reports, than they did to him.

General Augustus Henry Lane-Fox Pitt-Rivers is an unexpected, even unlikely, figure to play such an important part in the history of archaeology. It was a series of unlikely chances and unexpected deaths that gave him an enormous inheritance of landed property in the south of England. But when he became the owner of much of the area called Cranborne Chase in the downland of Wiltshire and Dorset, he found it was rich in antiquities of both Roman and pre-Roman date. He organized an army of diggers and trained assistants, an army not only in numbers but in discipline, an army uniformed in his personal colours, with his assistants allowed to ride on penny-farthing bicycles behind the General's dogcart as they went out to the day's 'campaign'. The combination of plenty of money, patient control and exacting discipline is one that is still all too rare in archaeology, so it is not surprising that the General achieved some remarkably successful results; he was digging between 1880 and 1900, yet his work seems to belong to the great excavations of the twentieth century both in size and in results and methods of recording. He was the first to uncover a whole village in such a way that the population could be estimated and food-supplies calculated. It was not until the 1930s that such work was repeated when the Germans opened up the entire village of Köln-Lindenthal, a Neolithic site near Cologne, and the value of excavation which showed farming techniques, population densities and food supplies was rediscovered.

The justification of the General's methods and care lies in the achievements of thousands of archaeologists who have learnt from him care and patience and the importance of almost everything they find. But it is more dramatically proved by the fact that the fifty-year-old records of his excavation of a village have been found comprehensive enough and satisfactory enough to support a new theory, developed within the last few years, about the economic trends of south-west England in Roman-British times.

The fusing of the three strands which went to form archaeology enabled Sir Leonard Woolley, the discoverer of Ur, who combined a career as one of the greatest of all archaeological diggers, with a positively unfair talent for brilliant, clear and lively writing, to comment in 1930:

> Until Schliemann dug at Mycenae and Sir Arthur Evans in Crete, no one guessed that there had been a Minoan civilization. Not a single written word has been found to tell of it, yet we can trace the rise and fall of the ancient Minoan power, can see again the splendours of the palace of Minos, and imagine how life was lived alike there and in the crowded houses of the humbler folk. The whole history of Egypt has been recovered by archaeological work, and that in astonishing detail; I suppose

we know more about ordinary life in Egypt in the fourteenth century before Christ than we do about that of England in the fourteenth century A.D. To the spade we owe our knowledge of the Sumerians and the Hittites, great empires whose very existence had been forgotten, and in the case of other ancient peoples, the Babylonians and the Assyrians, the dry bones of previously known facts have had life breathed into them by the excavation of buried sites. It is a fine list of achievements, and it might be greatly expanded. All over Europe, in Central America, in China and in Turkestan, excavation is supplementing our knowledge and adding new vistas to our outlook over man's past; and to what is all this due? Not to the mere fact that antique objects have been dug out of the ground, but to their having been dug scientifically.

This is an impressive statement of the achievements of the science of archaeology, which was, at that time, as Woolley points out, only seventy years old. But the importance lies as much in what is omitted. All the consideration is given to 'civilizations' and 'history'. There are but two mentions of 'the Stone Age' in the whole book, there is no use of the word 'Neolithic', there is but a passing reference to the antiquity of man or the origins of our species. The digging, so vividly described, is all among 'ruins'. There is no mention of the work in western or northern Europe or southern Africa which was employing the time of an equally large force of equally genuine archaeologists. For the truth is that Woolley himself was a classical humanist in the style that it was the pride of Oxford and Cambridge, the ancient English universities, to produce. His archaeology, professionally excellent, was in the tradition of the British 'travellers' and explorers of the Middle East. His aims, though higher than Layard's achievements, were fundamentally the same. And above all his appeal to the public, through his writings, was the appeal of the man who has found treasure. He cannot be blamed for this in any way. The 1920s and 1930s were the climax of archaeological treasure-finding (though not of treasure-hunting). Woolley's own magnificent finds at Ur – the Standard of Ur and the queen's head-dress for example – are now among the glories of the British Museum in London and crowds continue to flock to view them, crowds of children, students and ordinary people as well as scholars, as any visit on an ordinary Sunday afternoon will show. These were the years, too, of the discovery of Tutankhamun's tomb in Egypt, with its fabulous treasures of gold.

Woolley was far too intelligent not to have been aware of this. He explains the difference between the archaeologist and the treasure-hunter very carefully.

> In its essence 'Field Archaeology' is the application of scientific method to the excavation of ancient objects, and it is based on the theory that the historical value of an object depends not so much on the nature of the object itself as on its associations, which only scientific excavation can detect. The casual digger and the plunderer aim at getting something of artistic or commercial value, and there their interest stops. The archaeologist, being after all human, does enjoy finding rare and beautiful objects, but wants to know all about them, and in any case prefers the acquisition of knowledge to that of things; for him digging consists very largely in observation, recording and interpretation. There is all the difference in the world between the purpose and the methods of the scientific worker and those of the robber; it remains to be seen whether there is a corresponding difference in the value of the work done. Really the whole of this book is an attempt to explain the means which the archaeologist adopts, and to prove that they do achieve the end in view.[2]

The object of *this* book is to show how much and in what directions scientific archaeology has moved since Woolley wrote in 1930.

It was exactly in those years when Woolley was digging and writing that the next major developments in scientific archaeology were in their formative stage. The man most responsible was probably the most influential single personality that prehistoric studies and archaeology have yet known, Gordon Childe. He was an Australian who came to Europe early in his life. He was originally a Marxist, and, though he became more and more of a 'deviationist' as his life went on, his original intellectual attitude was at the root of many of his most valuable contributions to knowledge and ways of thinking. For one thing, he could read and understand twelve languages, and thus was able to use the work of Russian and East European archaeologists as well as those of western Europe and America. This gave him a unique ability to grasp a picture which spread from Turkestan, past the Caspian Sea, across south-west Asia, into the Middle East, and up the Danube to the coastlines of western Europe.

But more important, he saw human history, or prehistory, primarily as a history of technology. Instead of the rise and fall of civilizations, Childe interested himself in the rise and fall of technologies – the technologies of hunting and weapon-making, of herding and domesticating animals, of crop-growing and agriculture, of pottery and metalworking. In Jacquetta Hawkes' phrase, 'What he did, and did brilliantly, was to use archaeology to show man as a self-sufficient organism, specializing in brain, who has pulled himself up by the bootstraps of technology.' His most famous and influential book, *What Happened in History*, is a direct attack on the approach of most of the historians of his time. Childe says that what matters, what should be studied, is not the palace revolutions that replaced Pharaoh X with Pharaoh Y, nor the battles that enabled the Kassites to overcome the Babylonians or vice versa, but the technologies, the techniques that enabled people to produce such a surplus of food in the valley of the Tigris or Euphrates or Nile that great states could be set up.

It was Childe, too, who coined the term 'Neolithic Revolution' for the most important thing to happen in the entire history of our species, the discovery and development of agriculture with the associated development of the settled community – the village and later the city. The concept of Neolithic Revolution as an event has been attacked and discredited by archaeologists and prehistorians ever since Childe produced it. Technically and factually they have been right to do so, for the evidence is clear that there was no single event. Agriculture can exist without permanent settlements; deliberate and regular food gathering can occur without agriculture, settlement does not necessarily depend on agriculture and so on. Nevertheless, Childe's basic idea, with later modifications, seems correctly to have identified the most critical series of steps that have led to our modern world and man's position in it. It has also had enormous intellectual impact – men as far removed from archaeology as Hermann Kahn, the designer of nuclear-age 'scenarios' involving calculated numbers of 'mega-deaths', refer to the invention of agriculture as the most important single development in human history. The term 'Neolithic Revolution' is also exceptionally useful as a portmanteau term – to carry the load of ideas involved in the start of agriculture and the beginning of settled community life, and allow any particular prehistorian to take out what few items he wants to exclude.

The essence of Childe's theories was that the Neolithic Revolution had taken place in the 'Fertile Crescent', somewhere in the comparatively well watered, comparatively cool mountain slopes that stretch from Palestine, north through Syria and then eastwards through southern Turkey and northern Iraq (roughly following the northern side of the sweep of the two rivers Tigris and Euphrates) before turning south through the high country that divides Mesopotamia from the high plateaux of Persia (Iran). From some nucleus in this area the Neolithic Revolution had spread southwards to pre-dynastic Egypt, westwards up the Danube and across the eastern end of the Mediterranean, and eastwards to India. Where previously men had been hunter-gatherers, living only on what animal or plant food they could find, the ideas of the Neolithic Revolution diffused outwards from the Fertile Crescent and so did the necessary technologies, either by conquest or by the example of having regular and larger supplies of food available. Later evidence has shown that much of the detail is at least incomplete – the shape of the Fertile Crescent and the areas included in it have had to be changed and so on. But the idea of the Neolithic Revolution was enormously important, indeed much of the evidence that has led to its modification would not have been sought or found but for the impact of the original concept.

Yet, in many senses, Childe's ideas can be seen as the culmination of the work of those archaeologists who did not explore the rich ruins and graves of the Middle East, but who followed up the work of Boucher de Perthes and Worsaae in France and north-western Europe. This was the branch of archaeology that was formed by the two strands of those who were scientifically trying to prove the antiquity of man and the local antiquarians who wanted to know the history and prehistory of their own countries. In the broadest sense they had shown, not only their original discovery that man had existed alongside extinct animals, hunting and killing them with stone weapons, but that even within the Stone Age there had been development. They could show that the stone tools had developed from very primitive, simple, instruments into a wide range of highly refined tools and weapons. They had been able to distinguish between entirely different sets of stone tools, made with different techniques and therefore made presumably by different 'peoples'. They had built up a picture of successive cultures each more advanced than the one that preceded it – Mousterian, Chatelperronian, Aurignacian, Gravettian, Solutrian and Magdalenian, are the names of the cultures in the great French sequence. Most of the basic work was done in France, and it proved possible to establish this sequence of human development almost certainly because central France was just beyond the range of the southernmost extensions of the great ice sheets of the Glacial Ages. On the one hand the absence of ice allowed men to inhabit the area continuously and on the other hand there has been no destruction of sites by ice. But examples of these same 'cultures' – tools made in the same way, by the same techniques, with the same shapes – have been found from Uzbekistan to north Africa, and several of these cultures are believed to have entered France from the east or from the south. Remarkably little digging of Stone Age sites in the Middle East had been done by Childe's time, but it was clear that the cultures found in western Europe and France had been there too.

Possibly even more important in arousing intellectual interest in the men of the Stone Age had been the discovery of the cave paintings in France and Spain which had occurred around the start of the twentieth century, though further examples continued to be found throughout the first half of the century. These superb works of art, when they were finally proved to be the work of 'cavemen', showed that our Stone Age ancestors had been both artists and people with some sort of religious belief; the cavemen had indeed been human beings like us, with the same sort of intellectual equipment and the same powers of non-materialistic thought.

It was also shown that there had been types of men other than modern *homo sapiens*, such as Neanderthal Man; and there had been a shock to modern susceptibilities in the demonstration that Cro-Magnon Men had larger brain cases than we have. The building up of this picture of Stone Age man and his activities had been a slow and usually undramatic process, though much of it had in essence been achieved in the very first years of archaeology. The Great Exhibition in London in 1851, which was the first attempt to show a vast mass of people the scope of man's activity, with great displays of crafts and machinery as well as of arts, had contained no reference whatsoever to prehistory. The next great exhibition, the Universal Exhibition in Paris in 1867, admitted, not only the antiquity of man, but the fact that man had existed before and beyond the classical civilizations of Greece and Rome and the Bible. It was also the occasion when the new word 'prehistory' was first brought to the notice of the public at large. Most writers on the history of archaeology are agreed that the exhibition in Paris in 1867 marked the moment when the new sciences of archaeology and prehistory were officially admitted to be of age.

The exhibition not only showed the crude tools of stone which had enabled Boucher de Perthes to prove his point. It also showed that the French specialists, at least, had grasped that there was change and development within the Stone Age. The exhibition divided the tools of worked flint from the specimens of polished flint, and thus divided the Stone Age into two, the Old Stone Age and the New Stone Age. But in addition to this the Swiss were able to mount an exhibit of their recently discovered 'lake villages' and the Danes showed specimens from the first 'kitchen midden' site of Meilgaard. The general tone of this particular line of archaeology was set by the time of this exhibition in 1867, and the only major intellectual change involved in the next seventy years was the admission that the cave paintings and the early figurines (such as the 'Venus of Willendorf' and many others) showed that Stone Age man was both religious and artistic.

Although archaeologists regard the Universal Exhibition of 1867 as such an important date in their professional history, it is perhaps significant that the centre of the entire 'palace' of the exhibition was occupied by a colossal statue of the builder of one of the Pyramids, Chephren, a statue which the Egyptian government had loaned.

It was Childe's achievement that he provided the link, in theory at least, between Stone Age Man, artist though he may have been, and the sudden appearance of what we recognize as civilization, the start of our own world, in the Middle East, in Mesopotamia. And it was the idea of the Neolithic Revolution that filled this gap.

Under the inspiration (or goading) of Childe's theories, expeditions to great Middle Eastern sites dug deeper than the 'ruins' and discovered the villages of simple farming communities still further down. These were Neolithic peoples, using fine stone tools, growing crops, keeping domestic animals, and making very beautiful pottery. Two Middle Eastern sites can be taken as typical of this further penetration into the past: the Chicago University Oriental Institute's dig at Alishar near the modern Turkish capital of Ankara, where persistent digging took the archaeologists right through the levels of the Hittite civilization to Neolithic remains below; and further south in Cilicia, the Nielsen expedition at Mersin found a farming community which had to be dated to the sixth millennium B.C. or even earlier. Childe himself built up a picture of these Neolithic people spreading westward with their farming culture until they reached Spain, France and Britain, where they built the megalithic tombs and monuments such as Stonehenge and Carnac which had for so long been the hunting-ground of the antiquarians.

But it was not till after the Second World War that a major exploration of the Neolithic Revolution could be mounted in the Middle East. The digging of Professor Robert Braidwood (again from Chicago) and his wife, at Jarmo, a rather modest-looking village site in northern Iraq, and the excavation of Jericho in the valley of the Jordan by Dr Kathleen Kenyon of the British Palestine Exploration Fund, together caused one of the biggest upheavals in archaeological thought in the short history of the subject. They showed, sure enough, that agriculture must almost certainly have begun in this area, but they also showed that farming had a far longer history than anyone could have imagined. At Jarmo it was clear there had been farmers, who could not make pottery, back as far as 7000 B.C. The site also gave the first clues as to the nature of the wheat grown in those times, and started off the great hunt (which still goes on today) to find out just how man domesticated wheat and other plants, and how the plants, and man, evolved into their present mutual dependence.

Jericho confirmed the finding at Jarmo that people had been cultivating crops before they could make pottery, or, in other words, that what archaeologists called 'Neolithic' peoples were the *result* of the Neolithic Revolution and not the authors of it. It also showed that the Neolithic Revolution must have been a very long-drawn out affair – a whole series of different technological revolutions in fact. But Jericho's biggest surprise was the demonstration that cultivators who did not even know how to make pottery had built round houses of brick and were also capable of building an immense stone defensive wall (still more than thirty feet high when disinterred from the remains of later civilizations) and a stone tower inside the wall. These Neolithic men could, and did, build a town. The close relation of settled community life with the start of agriculture was called into question. 'The findings at Jericho . . . may indeed be said to be the most significant contributions to the history of *Homo sapiens* made since the last war,' writes Jacquetta Hawkes.

But underneath even the lowest Neolithic mudhouses in Jericho there were traces of occasional visits to the oasis by hunting groups of men who had weapons made in what archaeologists called 'Mesolithic style' – 'Mesolithic' meaning Middle Stone Age where 'Neolithic' means New Stone Age. These Mesolithic stone implements appeared very similar to 'Natufian' tools – the remains of a people which had been found in the caves of Mount Carmel, not very far from Jericho, some twenty years earlier by Professor Dorothy Garrod. Professor Garrod had also found the tools and skeletons of Neanderthal men on Mount Carmel and so her digging was well known and had been something of a landmark in its own time. But above Neanderthal men she had identified a group she called 'Natufians', people whose tools were Mesolithic in shape and manufacture, but where some of the stone cutting blades showed the telltale 'polish', or sheen, which implies that they have been used for the regular reaping of grasses or corns. (This polish is actually a deposit of the silica-based particles which, though minute, are one of the many components of grass stems.) It seemed that the Natufians had been regularly reaping wild grasses or plants, for there were no other signs of agricultural activities among their remains, and it seemed therefore that they were a people on the fringe between a pure hunting culture and the beginning of incipient agricultural development.

Finding traces of Natufian settlement at Jericho directly under the first dwellings of settled Neolithic people therefore linked the two 'Ages', Middle Stone Age and New Stone Age, and implied (though it has not yet been firmly proved) that hunters could settle down and quickly adopt farming practices and community life which had been developed elsewhere, though probably not so far away.

It also seems to bring the different strands of archaeology together; on one side extending even further back through time the steady increase of knowledge of the early civilizations of the Middle East and, on the other, the development of the understanding of early man, largely derived from studies of the stone implements of western Europe.

These discoveries had also pushed back the time at which we must estimate that farming began. But how far back? No one could say anything firmer than that farming had been practised by the seventh millennium B.C., and that was a great deal earlier than had been expected, since the very first civilizations of Sumeria, preceding even Ur, could not be earlier than 4000 B.C.

A whole variety of other questions have emerged in these last few paragraphs, questions of a sort which archaeologists were not accustomed to asking, still less answering. From what plants had modern crop plants, such as wheat, been developed and how? How had animals been domesticated? When and where and how had they changed under the influence of domestication? How, when and where was pottery-making discovered if it definitely came after the beginning of agriculture? Where had agriculture started? When had man actually begun to plant crops rather than just collect wild plants?

These were all questions which mere digging, recording and stratification could not possibly answer. New techniques were needed. Only plant geneticists could answer some of the questions archaeology was raising. Botanists and biologists and animal physiologists were plainly needed for others.

But it was from the physics laboratories that the first important new technique came – radiocarbon dating, to answer the questions beginning with 'when?' One of the first tasks given to radiocarbon dating after it had been established was to find the dates of the relics from Jarmo and Jericho. It silenced the arguments of those who would not believe that Jericho and Jarmo went back as far as their excavators claimed. It did more. It pushed the dates even further back. Jericho must have been founded, according to those first radiocarbon dates, as early as 7000 B.C. And farming therefore must have started even earlier. (In the light of the recalibration of radiocarbon dates in the early 1970s, it now seems likely that Jericho may date back to 8000 B.C.)

A good round date for the beginning of farming might well be 10,000 B.C. Or perhaps it might be safer to say 10,000 years before our own time. But the question of the date of the beginning of farming could only be raised, and can certainly only be answered when the techniques of other scientific disciplines, the technologies of the laboratory, are applied to archaeology.

1. J. Hawkes, *The World of the Past*, pp. 36–7.
2. L. Woolley, *Digging up the Past*, pp. 23–4.

2

The Importance of Dating

Possibly the most unfortunate man in the history of archaeology is James Ussher, who was born in Dublin in 1581. He was one of the great scholars of his time, he became Archbishop of Armagh and Primate of the Church of Ireland. Furthermore he was such a respected political figure in the troubled era of the Civil War in Britain that Oliver Cromwell allowed him to be buried in Westminster Abbey in 1656 according to the rites of the Church of England, which rites were otherwise forbidden in this period of the Protectorate when Presbyterians and Puritans held power. Yet it is not for this that Archbishop Ussher is remembered. It was his misfortune to have produced, in one of his scholarly works, the clearest and best-argued proof that the date for the creation of the world was 4004 B.C. Because some unknown person wrote this date – for which Archbishop Ussher was the authority – in a printer's official reference copy of the Authorized Version of the Bible, the material was included in later printings of the Bible. So for generations of English-speaking people, the date of the Creation at 4004 B.C. and Archbishop Ussher's name were linked, and it is always poor Archbishop Ussher who is now mentioned as the author of this date and held up to ridicule.

In fact Ussher held only what most men of his time, and generations before and afterwards, believed. Sir Walter Raleigh and William Shakespeare (in *As You Like It*) both mention this sort of date for the creation of the world, and other clerical scholars after Ussher refined the calculations to give a precise hour on a named calendar date. None of them should be ridiculed. Ussher and the rest were using the only evidence they believed was available about the past of the human race – the Bible.

The proof, by archaeology, that the human race was involved in Darwin's 'evolution' as much as the animals, implied that man had a history of untold thousands of years, but that this was but a portion of the time-span of the history of the world. Yet, when Ussher's date had to be abandoned, there was no real attempt for a hundred years to replace it with a better substitute. Man had been proved to have existed for thousands upon thousands of years yet there seemed no way of making more than the merest guess at the date of any event for which there was not some written record. Indeed there was a period when the concept of time was barely applied to prehistory. Geoffrey Bibby writes about the achievement of Christian Jurgenssen Thomsen, the 'father of prehistory' who first divided antiquities into the Three Ages of Stone, Bronze and Iron:

> In the verdict of history Thomsen's significance is greater even than his introduction of the Three Ages. He had introduced the idea of Time into prehistory. Whether his theory was accepted or denied, no one . . . could any longer suggest or believe that the time before the dawn of written history was a single, short, homogeneous period . . . That Thomsen produced the right answer is not so important as that he asked the right question. In future whenever two artifacts came together a museum curator was bound to speculate which was the older. And the whole of our present knowledge of the prehistory of mankind has come from the necessity of answering that question.[1]

But the question 'which was the older?' is not the same as the question 'how old is it?' The first question receives an answer in terms of relative dating; the second demands an answer in terms of absolute date. For many years the archaeologist usually had to be satisfied, and usually was satisfied, with a relative date.

The basis for relative dating, in archaeology, is simple. If one article is found above another (i.e. nearer the surface) then the first article, the higher of the two, is the more recent. Broadly speaking, if articles are dropped, or thrown out, or cast away, the one that is abandoned today will land on top of the one that was thrown away yesterday. To write in terms of single articles is obviously an oversimplification, but this is the basic principle of stratification, a principle that archaeology inherited from one of its progenitor sciences, geology. In practice stratification must have developed more on these lines – a tribe of people occupied a cave regularly during their summer hunting for a hundred years, each year they dropped animal bones and broken stone tools on the floor of the cave; then they were defeated in battle, or the climate changed and the animals moved away, and so they stopped using the cave every summer; for a hundred or a thousand years the cave was unused and a steady drift of sand, or dust, or earth and rocks from the roof of the cave covered the debris left by those men. Again the climate changed or the fortunes of battle reversed and another group of men with different stone tools and different habits started to occupy the cave regularly every winter as they herded reindeer; they, too, left a layer of debris and the debris again was covered, until the modern archaeologist comes along with his spade and trowel. What he sees, as he digs down through earth and debris and earth again and more debris, is that those people who left the topmost layer of debris must have used the cave after the people who left the lower layer of debris. Sometimes there will not be a convenient dividing layer of empty, 'sterile', soil or sand, and one layer of debris will lie directly on top of the first. Sometimes a change of climate will deposit a different type of earth over a layer of debris so that the change is clearer than ever. There may be confusion caused for the archaeologist by the second set of people digging graves for their dead in the floor of the cave so that the bones and grave-goods of their dead lie among the remains of those who had not visited the cave for a thousand years past. But the principle and the basis of relative dating are clear.

In the context of the Middle East, Woolley provides a fascinating description of how ruin and stratification take place in the 'tells', the great mounds which signify places of human habitation.

The commonest building material is mud brick, and mud brick walls have to be thick; when they collapse the amounts of debris is very great, and fills the rooms to a considerable height, and as you cannot use mud bricks twice over, and the carting away of rubbish is expensive, the simplest course is to level the surface of the ruins and build on top of them – which has the further advantage that it raises your new mud-brick building out of reach of the damp. In Syria and in Iraq every village stands on a mound of its own making, and the ruins of an ancient city may rise a hundred feet above the plain, the whole of that hundred feet being composed of superimposed remains of houses, each represented by the foot or so of standing wall which the collapse of the upper part buried and protected from destruction.[2]

So when the archaeologist comes to dig down through the remains and ruins that constitute a 'tell', or mound, he may easily find twenty distinct occupation levels – each one distinguishable from those above or below it, each one representing a distinct culture or civilization.

Sometimes one level of occupation is clearly marked off from the next by a layer of ash or blackened material – the city has been destroyed by fire. To the archaeologist this clear stratification is most valuable. Woolley explains why:

If the field archaeologist had his will, every ancient capital would have been overwhelmed by the ashes of a conveniently adjacent volcano. It is with green jealousy that the worker on other sites visits Pompeii and sees the marvellous preservation of its buildings, the houses standing up to the second floor, the frescoes on the walls and all the furniture and household objects still in their places as the owners left them as they fled from disaster. Failing a volcano the best thing that can happen to a city, archaeologically speaking, is that it should be sacked and very thoroughly burnt by an enemy. The owners are not in a position to carry anything away, and the plunderers are only out after objects intrinsically valuable, the fire will destroy much, but by no means everything and will bring down on the top of what does remain so much in the way of ashes and broken brickwork that the survivors, if there are any, will not trouble to dig down into the ruins; a burnt site is generally a site undisturbed. It is where cities have decayed slowly that least is to be found in their ruins; the impoverished inhabitants will have pulled down the older buildings to re-use the material in their own hovels, they will make nothing good of their own and they will certainly leave nothing behind them when at last they desert the place; the top levels of such a site generally produce therefore few objects, and not much history except the melancholy history of decadence.[3]

On a great site like that of the city of Troy the archaeologist will find some layers where fire and pillage have both destroyed and preserved in the way Woolley describes and other layers where the city of the time decayed slowly until it was deserted, perhaps for several hundred years until some new people came and built upon the dust covered ruins probably valuing the mound of human debris only for its defensive possibilities.

When it comes to the modern archaeological excavation of a rich cave site the stratification has to be even more carefully investigated. Consider the description by Carleton Coon of the first steps in digging out his famous find at Belt Cave on the shores of the Caspian Sea. He was digging twenty years after Woolley and he wrote:

As these men took out the first twenty centimetre strip, I realised that we faced a complicated problem in stratigraphy, because this thin strip showed a variety of earths from one end of the trench to the other. At the front end, black soil that had drifted in from a bank outside encroached some twenty centimetres over our excavating area. Moving backward we found this to be followed by a metre more or less, of yellowish soil, which turned out to be a lens – a thin discontinuous layer of intrusive material. Beyond this we came upon a fine greyish soil, composed of a mixture of brown earth and wood ash. In it the blackened remains of many hearths were visible. In the back half of the trench, some five to ten centimetres of peated manure still overlay this grey earth. The grey earth was pay dirt. It contained a wealth of flint, bone and pottery, including twenty-three flint blades of the finest and most delicate quality, and of Neolithic type.[4]

Carleton Coon had found Neolithic remains at the top of his cave deposit. As he dug down further and further he laid bare more than twenty different levels. At the bottom was a layer of clay laid down by the Caspian Sea itself when it had been much higher than its present level. The first inhabitants of the cave as sea-level fell were Mesolithic – Middle Stone Age – men who hunted seals along the shore. When they left the cave was untenanted for some three thousand years, until an entirely different, but still Mesolithic, group came along who lived by hunting gazelles which now abounded in the area as the steppe-land had developed. Above their remains were Neolithic cultures which had not yet discovered how to make pottery, which did not seem to practice much, if any, agriculture, but which plainly herded and ate sheep and goats. From then on upwards the levels showed the domestication of cows and pigs, the development of grain-reaping, and the manufacture of pottery, till the top layers, the first to be found, contained the remains of a full Neolithic culture. Incidentally Carleton Coon does not claim that this site represents the place where pottery was invented – for when it first appears in his record it is of quite a high quality – nor that the inhabitants of Belt Cave were the first farmers or herdsmen. (Later his materials were dated by radiocarbon which placed Belt Cave's first inhabitants, the seal hunters, at 9500 B.C. and its final Neolithic owners at 5330 B.C.)

In the same description of his work at Belt Cave, Carleton Coon also gives a very clear explanation of the confusions that can beset even the most careful study of stratification.

Level Number Eight was entirely without pottery, and we found no more in this trench except one spook in level No. 10. [Author: Coon gives his levels increasing numbers as he goes down – 10 is therefore deeper than 8.] A spook is a potsherd or other specimen that is found in an apparently undisturbed level in which it does not rightfully belong. Its presence can be explained in several ways; by the action of roots – and roots went down four metres in this cave; by falling down the hole of a prehistoric burrowing animal such as a mole, vole, or small rodent; or by slipping down the rock wall of the side of the cave, where air spaces sometimes occur probably owing to the action of water. At any rate the presence of spooks is interpreted statistically. One or two may be forgiven for being in the wrong place, but if too many turn up, then either they indicate a disturbance as where someone has dug a grave through several levels, or else they really belong in the place where they are found and must be interpreted as trade objects or part of the local toolkit of that period.[5]

Despite the presence of spooks – anomalies is a more technical word for hiding the fact that a bit of evidence is embarrassing –

relative dating by stratification has proved highly successful. But it only tells the excavator the order in which things happened in his particular cave or Middle Eastern mound. It does not tell him when exactly it all happened in the long course of history and prehistory. It does not give him an absolute date in terms of years B.C. or A.D. It does not even tell him whether the deposits in his cave are older or younger than the deposits in the next cave he investigates.

It has proved perfectly possible, however, to extend relative dating from one site, cave or 'tell', to another. If cave (or tell) number one has three layers, A, B and C, going upwards, and if in cave number two the lowest level contains tools of stone and pieces of pottery which are very similar to those in layer C of cave number one, then everything above the bottom layer in cave number two is later than all the contents of cave number one. When this sort of principle is applied by scholars and careful investigators to a broad area like the Middle East it is perfectly possible to build up a great structure of logical relationships and a system of relative datings which is internally consistent and which can spread over a wide area.

Pottery, the commonest single type of debris left behind by human beings, which can survive the erosion of time and weather and bacteria, has always provided the strongest links and clues which have enabled archaeologists and prehistorians to build up an ordered time-sequence of past events and places. Pottery, furthermore, is made by individuals – craftsmen, or in many cases craftswomen. There are therefore styles and traditions in the shape of pots, in the decoration on them, in the materials of which they were made which provide identifying strands running through centuries of time and which the archaeologist can use to pick his way among the political and cultural movements of peoples and traditions. It is possible to follow even the process of the development of a particular line of pottery as it becomes steadily better in material and technique, steadily more beautiful or more elaborate in style. This relative dating can be carried even further – for a level in one 'tell' may contain pottery of a certain type which is plainly a more developed, and therefore later, version of pottery found elsewhere.

There are limits, however, to a system of relative dating of this sort, where the pottery and artefacts in different layers at many sites are cross-related to give a 'time-structure' over any large area. First, the whole system is only as strong as the weakest link in the chain of argument. Any large system of evidence and argument will eventually have a number of weak links, any one of which could lead to errors affecting all the other arguments and conclusions. For instance it is likely, but not necessarily true, that there must be a direct human connection between two distinct places where the same sort of pottery is found. It is conceivable, however, that two quite unconnected human groups might make the same sort of pottery if both groups had the same needs and the same raw materials. Or it is possible that the two similar types of pottery might represent similar answers to a single set of problems with a large time-gap between them. (There are examples in the animal world of what is called 'convergent evolution', where two different species develop the same sort of answer to a problem posed by the environment; and in modern engineering developments we can see remarkably similar answers developed by different groups of men faced with the same set of problems (the shape of supersonic aeroplanes is a good example).)

When archaeologists, like detectives reconstructing the picture of the past from the clues left in the present, conclude that one type of pottery is later in date than another because one sample seems to be 'finer', or 'more artistic' or 'more sophisticated' or 'better-glazed', then the possibility of errors becomes greater. Such judgements are necessarily subjective – what seems 'finer' to us may well have appeared the opposite to men of earlier cultures. And even if we accept the judgement of the excavator or art-historian about the relative qualities of the two samples, we know that cultures, or civilizations, or artistic traditions, can degenerate as well as advance. Hence it is quite possible that some samples of poor-quality pottery in a certain stylistic tradition might be later in date than better-quality examples because the culture, or the materials available, are degenerating rather than improving. Thus, to the outsider at any rate, some of the enormous intellectual constructs created by archaeologists and prehistorians, which allow them to compare the relative dates of objects found as far apart as the southern exit to the Persian Gulf and the Turkish Black Sea coast on stylistic evidence alone, seem to be of doubtful validity – we can suspect that there may be so many minor errors in the long chain of connections that the conclusions are worthless.

The art-historian using stylistic evidence to establish trading, or human, connections between distant sites, or to determine the relative dating of different occupation-levels, has, however, become an important figure in archaeology. Many modern archaeologists and prehistorians, especially those with a leaning towards scientific methods, feel that he has become far too important. It is not only that such methods can lead to mistakes and wrong deductions – anybody can be forgiven for this – but that the scientific archaeologist feels that the profession has been led, in some cases, into blind alleys and wrong attitudes even in its basic tasks of excavation. The use of 'stylistic' evidence, especially based on pottery finds, has undoubtedly had great success in sorting out the problems of Middle Eastern chronology. But it is felt by some that the application of the 'subjective' thinking involved has been wrongly extended to the analysis of stone and flint tools in many of the most productive French sites. Here the search has often been for 'typical' flint tools; a search for the one precise sort of tool which will enable the archaeologist to identify the whole level he is excavating as belonging to one specific culture, so that he can unhesitatingly classify the whole assembly of stones unequivocally as 'Solutrian' or 'Aurignacian'. It is claimed that this search, discarding large masses of stone tools because they were not 'typical', has led to failures in interpretation of what the actual masses of tools have meant, or even to wrong methods of excavation in which the wrong things have been looked for and recorded.

But even if the relative dating system has been worked out correctly to connect two distant sites, the archaeologist is faced with another, quite different problem. It is simply the fact that human beings have not necessarily progressed at the same rate in different parts of the world. This means at the extreme that stone arrow-heads found in North America cannot be given any relative dating by comparing them with rather similar arrow-heads found in Europe. More realistically, it is clear that the Neolithic Revolution did not occur in western Europe till two thousand or more years after it took place in south-west Asia. The idea of agriculture may have spread west from the Fertile Crescent by conquest, or by example, or by trade, but a fully agricultural Neolithic site in Denmark or Britain must obviously be much later than a similar site in Iraq, though more than this it is impossible to say.

The problem can be presented in an even more striking way. For some peoples of the human race the Neolithic still exists today. The tribes still being discovered in the highlands of New Guinea are still Stone Age Men. Any one of their villages provides us with a Neolithic site dated approximately 1970 A.D.

In strictly archaeological terms the real problem this poses is that we cannot say with certainty which way things moved, in many cases. For instance the stone tools described as 'Gravettian' have been found in places from the Russian steppes to the foothills of the Pyrenees. The modern archaeologists question whether there can really have been one people covering this enormous range, one 'Gravettian empire'. But in that case one may ask whether Gravettian tools, however they spread, began in the East and travelled west, or vice versa, or began in the middle and spread outwards. And from this it follows that two dates which are the same by relative methods – that is, are Gravettian – will be different in absolute terms.

Finally relative dating can be useless, in the sense of being uninformative about important matters, if connections cannot be made between one site and another. It has proved possible to set up links, albeit rather flimsy ones, between the general system of relative dating in the Middle East and the Indian civilizations of the Indus valley, the Harappan culture. But no connecting link of any sort can really be found with ancient Chinese beginnings of civilization. And still less is there any valid connection between Middle Eastern and European early civilizations and the impressive ruins of Central America. It has even been suggested that the Chinese might have received the idea of agriculture literally by 'word of mouth' from Iraq through India or Persia, but there is no shred of real evidence for this and therefore no relative dating between the start of agriculture in south-west Asia and in China.

The weakness of relative dating, the necessity of finding some absolute method of dating prehistoric finds, and the problems of stylistic interpretations can be summed up in one famous case – the problem of the Megalithic tombs of Spain and Portugal. This particular illustration also has additional significance because its solution, in the last two years, has led to the most recent 'revolution' in archaeology and has been instrumental in bringing about the birth of the 'new archaeology'.

As recently as 1958, only fifteen years ago, one of the most distinguished British prehistorians of his generation (and, like Woolley, a brilliant writer and clear and lucid popularizer) Glyn Daniel, recently appointed Professor of Archaeology at Cambridge University, published his book *The Megalith Builders of Western Europe*. It concerns the great series of monuments and graves – the dolmens, menhirs and barrows, long or round – which were among the first things to excite the interest of the old antiquarians. They are distributed, broadly, round the west coast of Europe, from Denmark in the north, through the British Isles, down the coasts of France, Portugal and Spain, and on into the Mediterranean, through the great prehistoric 'Temples of Malta' until a relationship can be seen between the gaunt and colossal upright stones on a Welsh mountainside and the great tombs of Mycenae in Greece, where Schliemann dug and believed he had found the death mask of Agamemnon. The critical point is the Iberian Peninsula, Spain and Portugal, where there are both the simplest single-chamber tombs to be found in the north and rich, dramatic, well-constructed funerary monuments in the south, some of which use building methods remarkably like the Mycenean tombs.

It is generally agreed that the idea of building such elaborate, often gigantic, structures, in which to bury the dead, must plainly have been a religious idea. Hence the spread of the concept of the megalithic tomb must have been something like the spread of a religion by conversions, whether forcible or by example.

The first theory about the Spanish and Portuguese tombs was the work of several archaeologists, British and German, but most of all by the Catalan, Bosch-Gimpera.

These archaeologists were, of course, much affected by the notion of evolutionary progress in the elaboration of tomb-types and in material culture; they therefore thought the 'Neolithic' grave goods of the North Portuguese single-chambers must necessarily be earlier than the Chalcolithic grave goods of the South Spanish and Portuguese tombs [Author: Chalcolithic describes cultures half-way between the New Stone Age and the Bronze Age]; and that 'simple' structures . . . of North Portugal and Galicia must of necessity be earlier than the 'complex' and 'elaborate' structures of Southern Iberia. . . . I have put some of these words in inverted commas because whenever writers speak of primitive and complex, simple and elaborate, in relation to megalithic tombs we are getting near the point when we should ask sharply whether these terms are being used subjectively to denote observer-imposed prejudices.

Bosch-Gimpera regarded the whole Iberian development as indigenous, he boldly derived all the Iberian chamber tombs from the single chambers of Galicia and North Portugal, and envisaged a dual spread from this north-western corner of Iberia, the one east along the north coast of Spain to the Basque-Catalan area, the second south and south-east to Alcala and Los Millares. As this tomb idea spreads southwards it becomes more complicated . . . Alcala and Los Millares then stand at the height of a magnificent local Iberian development – the full flowering of which is in the late Chalcolithic period and the beginnings of which are well back in the Neolithic. To put this in terms of absolute chronology, Bosch-Gimpera suggested that the north Portuguese and Galician single chambers date between 3000 and 2500 B.C. and Alcala and Los Millares after 2300 B.C.

This work reported by Glyn Daniel dated from the first decades of the twentieth century. But, in the years that followed, the resemblance between the finest of the south Spanish tombs and the great tombs of Mycenae grew more and more compelling to many archaeologists, until a compromise theory emerged which held that Bosch-Gimpera was right in general, but that the most magnificent tombs were a separate development which had spread from Greece westwards along the Mediterranean and 'met' the Spanish native tombs which were rather more primitive.

Then, in the early 1930s, the pendulum swung even further, largely under the influence of a group of British archaeologists, Forde, Fleure and Peake. (British classical education has tended to make English scholars extreme philhellenes.) Gordon Childe supported the new notion because it fitted so well into his theories. Glyn Daniel writes,

> From this new viewpoint it appeared that the single chambers of north Portugal and Galicia could equally well represent degenerate and impoverished, and not primitive and undeveloped cultures. These writers suggested that the earliest chamber tombs in Iberia were in fact the fine South Iberian corbelled Passage Graves [Author: i.e. the graves like Mycenean ones] and that these were translated into megalithic tombs in areas where suitable stone for dry-walling did not exist or when the complicated technique of corbel-vaulting was no longer practised or where it was not easy to cut tombs in the rock . . . This is a complete reversal of the classic Bosch-Gimpera point of view and it makes the 'primitive' Neolithic cultures of the early third millennium B.C. into impoverished Chalcolithic cultures of the middle or late second millennium B.C.

As if more than a thousand years' difference in dating was not enough to confuse the issue, there then came, in the 1940s and 1950s, a swing right back to the original position of origin. This view largely comes from George Leisner, and he sees the Iberian megalithic tombs as being a purely native tradition, but starting off in two different ways, from sources in central and south-east Spain, and mixing to develop into the full magnificence of the height of the culture. From this, it must be presumed, the idea spread eastwards along the Mediterranean to Greece. Presumably, too, the idea of megalithic graves had also spread northwards. But here there is a complication, for the archaeologists had also agreed that the idea of graves marked by enormous stones must also have had an independent origin in Denmark, from where it had, at least to some extent, spread southwards, only to meet an incoming north-easterly wave of people with religious ideas about burial who were able to take on the Danish megalithic building systems.

Glyn Daniel in 1958 made his own decision thus:

The two things to keep foremost in the mind are that while the use of large stones may occur independently in different places, collective burial in a chamber tomb is surely a complicated and religious idea, and that this religious idea is manifested not only in burial and funerary custom and architecture but in the delineation of a cult figure. To my way of thinking disputes about typology and tombtypes become academic exercises if we forget that the great megalithic tomb-builders of Western Europe were imbued by a religious faith, were devotees of a goddess whose face glares out from pot and phalange idol and the dark shadows of the tomb walls, whose image is twisted into the geometry of Portuguese schist plaques and the rich carvings of Gavrinnis and New Grange. The question turns on Megalithic Art. The distribution of Iberian megalithic art is a coastal one. The cult-figure is seen at its best and freshest in the dry-walled tombs of southern Iberia and appears less recognisable as we move north and west into Portugal and Galicia. What is more this figure has very strong parallels in the East Mediterranean and Near East, in contexts earlier than our earliest Spanish ones. I find it very difficult to believe that our Spanish goddess and our oculi motifs do not come from the early goddess figures of Cyprus, Crete, the Cyclades and western Anatolia.

And so he concludes,

Settlers arrived in southern Iberia from the East Mediterranean somewhere before 2000 B.C. bringing with them the custom of collective tomb burial and a strong religious belief in an Earth Mother Goddess. . . . From these centres in south and east Spain and perhaps also in the Tagus estuary the Iberian Chalcolithic culture developed and, as it developed, perhaps in contact with earlier aborigines who built megalithic cists, the Megalithic Passage Grave was evolved and out of these developments the tombs of Northern Portugal and Galicia and the Cordoba-Granada area some of which certainly date after 1500 B.C.[6]

But others still believed that the magical tradition or religious belief that led to the building of the tombs originated in western Europe or North Africa, or, more simply, that the Mycenaean Greek tombs were an importation from the west.

The point is that only dates, absolute dates, independent of relative cross-matchings, independent of subjective judgements about the megalithic art or whether there was development or degeneration, can settle the argument. It is not just an argument for priority of date for anyone's local or national culture. It is an argument about whether one of the most impressive religious ideas of prehistoric times moved from east to west or vice versa. If the idea moved from east to west then the theory that the cultures of west Europe had arrived from a nucleus in the Middle East even before Graeco-Roman times is supported strongly, and this is the idea of the diffusionists under Childe's banner. If the idea spread from west to east then our entire set of concepts about the beginnings of civilization will have to change.

The archaeologists, of course, knew quite well that they needed absolute dates – that what they wanted was some sort of clock mechanism which could be found at any site, or used at any site, and which would tell them what the time had been when that site was used. They needed a clock that would tell them the time in any part of the world, but naturally it had to be a clock that ran backwards, a clock that told them the 'time before now'.

At first the archaeologists' only hope of finding any absolute date came from written history. Greek history went back to about 500 B.C., but, more important, as excavation in Egypt proceeded, and more and more of the colossal number of inscriptions on tombs, temples, and buried coffins were discovered and deciphered, an accurate historical date system was built up for ancient Egypt stretching back to 3000 B.C. It must not be thought that this dating for ancient Egypt was 'accurate' in the sense that we use that word for the dates of battles in the American Civil War or even for the dates of Roman history. But, to the prehistorian in general, Egyptian dates were accurate – correct within twenty-five years at worst for most of the first and second millennia B.C., and still correct within a hundred years even at the very worst, to the start of the first dynasty some time just after 3000 B.C.

Any event or ruined city level which excavators discovered outside Egypt but which would give a relative date with anything in Egypt could therefore be given an absolute date, often of considerable accuracy. Discoveries such as the Royal Library of clay tablets covered with cuneiform writing in the Assyrian capital of Nineveh, provided, when the writing had been deciphered, an enormous structure of historical events concerning not only Assyria, but its neighbours, which were quite easily cross-related at several points to Egyptian records. Thus a dated structure for a much wider Middle Eastern history emerged. One of the most valuable excavations in this field has been the work of the French team, under Professor Claude Shaeffer, at the ancient seaport of Ras Shamra on the Syrian Coast. This town, nicknamed 'the Shanghai of the mid-second millennium B.C.', and named in earlier times as Ugarit, has provided a fantastic variety of trade and diplomatic 'documents' and records which have enabled a whole series of states and civilizations – such as the Hittites – to be firmly anchored into Egyptian chronology and therefore given absolute dates of their own. The detail is so great that individual letters and treaties between named Hittite monarchs and named and dated Egyptian pharaohs have been found. The first alphabetic script was also found at Ugarit.

The Babylonians and Akkadians, too, left written records, and these in turn have been cross-related into dated Egyptian history. With the Code of Hammurabi inscribed on stone as the most famous of their 'documents', these Babylonian records have allowed dated history to be pushed back beyond Egyptian times, though with ever-decreasing accuracy and certainty about the dates. This in turn allows the Sumerians to be tied into the dating system and so back to the cultures of Ur and al 'Ubaid. But before 4000 B.C. all accurate dating has ceased. History, in the sense of a dated time-sequence of recorded events, has ceased. Therefore, before 4000 B.C., there is no hope of obtaining an absolute date anywhere in the world by the use of the traditional methods of historian and archaeologist combined.

Furthermore, even this absolute dating based on Egyptian chronology covers only a very small portion of the world. It is, for instance, impossible to obtain any firm dates for early Greek history from the Egyptian source. There is but one reference to the 'peoples of the Sea' attempting an invasion of Egypt, and some experts believe this to refer to the Greeks or Minoans, since there is one cross-reference to a word that is rather like the Greeks' own name for themselves as 'Achaeans'. But this is slender evidence indeed. If any attempt at dating cultures further afield is attempted from a series of relative cross-datings, then the structure soon seems distinctly wobbly. More than twenty beads of blue faience, a typically Egyptian product, have, for instance, been dug up in Britain. Some attempt at dating can be made from these by saying that any layers in which they have been found cannot possibly have been earlier than the first appearance of such objects in Egypt. Indeed they must, presumably, have been much later, since the beads would have taken a long while to travel to Britain by whatever route of trading or gift-swapping one can imagine. Little of real value can be built upon such slender clues, and the latest work, in any case, seems to show that the faience beads are not Egyptian at all.

To speak of any system of absolute dating based on Egyptian chronology even going back as far as 3000 B.C. betrays a somewhat parochial, European-oriented thinking in any case; a thinking still not free from the shackles of the classical education in Graeco-Roman studies which biased so much of early archaeology. For virtually the whole of Asia, Africa, India and China, and all the Americas and Australia, dating prehistoric objects by reference to Egypt is meaningless. The little light shed by Egyptian records covers such a minute area of prehistory that it is virtually irrelevant to archaeology as a whole, and does nothing in any practical way to provide the much needed clock. Many efforts were made to find the clock-device needed by archaeologists.

It was in 1946 that Frederick Zeuner, Professor of Environmental

Archaeology in the University of London, really crystallized the problem, and the attempts that had been made to solve it, in his book, *Dating the Past.* He elevated the whole thing to a scientific discipline in its own right which he called 'geochronology', and he attempted to pull together all the scattered work that had been done into one huge system of dating the past. The book became a standard textbook and went through edition after edition with more and more evidence being added each time up to 1958. But in his first preface Zeuner was quite clear what he was doing: 'The chief field of application of geochronology is in prehistoric archaeology and human palaeontology. The evolution of man, both from the anthropological and the cultural points of view, cannot be understood properly unless the time element is introduced.' It is an ironical comment on archaeology that this had to be said ninety years after the subject had got onto its feet.

On the first page of his introduction Zeuner is blunt:

> History has the advantage of written records and of calendars which provide more or less reliable *dates* and therefore allow of estimating the duration of periods of evolution. For Prehistory no calendars are available. Up to not many years ago the time scales suggested for the evolution of early man and his cultures were pure guesses. From a scientific point of view they were worthless. Yet nobody can claim that the time problem in the evolution of man is of little significance.[7]

Zeuner claimed that, by the time he was writing, at least some method had been found for giving some sort of date for all the main periods of the earth's history. The first section of his book covers back to 1000 B.C. – early history and late prehistory; next there is the stretch back to the end of the Ice Age – the time of the Three Ages of Stone, Iron and Bronze; next the Ice Ages, which for men corresponded to the Old Stone Age, stretching back to about one million years; and finally he has a section on dating the history of the earth before man arrived, covering the rest of the evolution of life and starting at the formation of the earth 3,500 million years ago. For his first period, back to 1000 B.C., the only 'clock' he could quote as useful was tree-ring analysis, or dendrochronology. For pushing further back than that he relied first on 'varve chronology', and then on a combination of botanical (pollen analysis), biological, environmental and climatological studies. This took him back to about 15,000 years ago. Beyond that, covering the Old Stone Age, he relied on careful geological and climatological studies of the advance and retreat of the ice sheets, coupled with the hope that astronomical studies of the variations of solar radiation would have left some geological traces in areas such as Africa which had not been affected by ice.

Tree-ring analysis was given pride of place by Zeuner, at least in terms of time, and it was the only 'clock' he could suggest for recent prehistory. By chance it was the first 'clock' ever to be suggested, the original proposal having come from De Witt Clinton when he examined the earthworks at Canandaigua in New York State in 1811. Because of the enormously important achievement of tree-ring dating in the last couple of years, it will be the subject of a special chapter (Chapter 3). But, at the time Zeuner wrote, it had proved rather disappointing everywhere except in the unusual conditions of the south-west of the United States. Zeuner could only sum up in these terms:

> There are obviously good chances for applying tree-ring dating to other regions, especially temperate Europe, but progress will be slow and a good many years may elapse before reliable results are achieved even for the latest prehistoric periods, as these are earlier in Europe than in America. Yet tree-ring analysis may one day provide a help in dating historic objects in Europe as well as parts of the Mediterranean.[8]

Varve chronology began as a distinctly Scandinavian method of finding a clock, although it has since found uses and supporters in the U.S.A. Basically it depends on the steady melting of the ice sheets of the Ice Age. When a glacier melts under the summer sun the water that pours away is far from clear, in fact it is thick and whitish and is often called *Gletschermilch* – the milk of the glacier. The opacity of the water is caused by the vast quantities of small particles of sand and clay which have been picked up by the glacier as it formerly advanced across the land. When this suspension of small particles is dumped into a lake, or even a quiet river, the particles slowly settle towards the bottom, and as they do so they sort themselves out, the heavier particles falling first and the smallest and lightest landing on the bottom last. It may even need the freezing of the water in winter to get the smallest particles to the bottom. But the following spring and summer the same process is repeated; until eventually the bottom of the lake builds up in an enormous series of perfectly distinct layers. These layers are called *varves* in Swedish.

As long ago as 1878 Baron Gerard de Geer, a Swedish geologist/archaeologist was struck by the regularity of these varves, which are often exposed during construction or quarrying work. He soon realized, however, that there were also noticeable differences in the thickness and consequent patterning of the lines in the exposed clays. And once he had started studying the varying thicknesses of the varves he realized that similar patterns of variation were shown in entirely different places. The reason for the formation of the varves was already understood, but it was de Geer who realized that the patterning in the thickness of the varves must represent a pattern imposed by the varying weather of the sequence of summers that had melted the ice. A thick varve implied a large quantity of melt-water from the glaciers and therefore a long, hot summer, whilst a thin varve implied a cool, and humid summer during which less ice had melted from the glaciers.

He began counting the varves in various different places around Stockholm, allotting one year to each varve. By relating one series of varves to another at points where the patterns of varying thicknesses were identical, he gradually built up a chronology of varves covering many thousands of years. Thus to de Geer goes 'the credit of having designed the first scientific method of dating geological events in years' – the words are Zeuner's.

The problem at this stage was, then, to find out just which of the varve years corresponded to which calendar year: the varve chronology had to be linked to the present day in some way. This represented a lifetime's work – de Geer had started in 1878, he built up a school of followers and assistants, and was still publishing results in the 1940s. Indeed research work to strengthen the 'de Geer chronology' is still going on to this day.

It was known from other geological studies that the ice sheets had steadily retreated northwards across Sweden until they had finally broken up into two separate parts. This break-up was called the 'bipartition' and it was known that the gap between the sheets first appeared near the place now called Ragunda. Very near Ragunda a lake had been left behind, and historical records showed that this lake was accidentally emptied of water in 1796. From 1905 de Geer had been steadily working up from the south of Sweden, building up his varve chronology, which outlined the retreat of the ice in years. He had high hopes that the drained lake in Ragunda would solve the last of his problems and connect his series of dates to modern times. He was disappointed in this respect, but, nevertheless, the series of varves from the bed of the former lake was enormously valuable. They added three thousand years to his chronology and, in particular, they included one extremely thick varve which de Geer had also found in some of his other, more southerly, sections. He decided that this very thick varve must represent the great outpouring of the time when the ice sheet broke into two parts and, presumably, released waters from all sorts of lakes and ponds previously blocked up by ice. This varve – the bipartition varve – he chose as the zero point on his scale, with all later years up to the present marked with a plus sign.

It was left to one of his students and assistants, Liden, to link up de Geer's varve chronology to the present day. He did so when he found a series of varves near the mouth of the Angerman river in north central Sweden which ran from the topmost end of de Geer's series right into recent times. This was possible because this area of Sweden has risen steadily further and further above sea-level ever since the weight of the ice sheets was removed from the land. On the calculations of Liden, de Geer's zero mark, the year of the bipartition, became 6839 B.C. In theory any archaeological finds which could be related to near-by varves could now bed ated in precise years.

But it did not work out like that. Even before the linking of de Geer's chronology with modern times had been completed there were serious technical criticisms of his methods and the over-simplicity of counting one varve to each year. The Finnish archaeologist Sauramo, who followed de Geer in varve-counting and study till he was second only to the originator of the idea, did not accept a simple pattern of varying thicknesses as enough to link varves from one place with those of another. It was shown that varve thicknesses could vary from one place to another even around the same lake. Sauramo insisted that the sections of varves must be taken back to the laboratory and studied for colour, coarseness of grain, hygroscopicity and plasticity before a series of varves from one place could be correlated with those from another. It also became clear that varves could 'go missing' for a year, and that there were such things as 'winter layers' which could throw the varve-count out of true.

Nevertheless varve-counting throughout Scandinavia did enable a whole series of correlations to be made between the timing of various 'events' as the ice sheets receded and an acceptable chronology for these events in terms of years was built up. Academic disputes about precise dates continue and there may be differences of 2500 years over events that took place 12,000 years ago. But varve-counting achieved an accuracy, even so, that had been undreamt of before, and which could not be achieved anywhere else in the world.

'Anywhere else in the world' is a vital qualification, though. Herein lay the chief reason why varve-counting never managed to provide a 'clock' system outside a limited area in Scandinavia. It was not universal. It could not be made to apply to other parts of the world. The problem was in many ways the same as that of trying to link relative dating from stratigraphy through to Egyptian absolute chronology – the arguments just would not stand the strain. De Geer himself believed that it was possible to make 'teleconnections' between his Swedish varve-series and other series of varves found in many distant parts of the world. He believed he had found valid comparisons with Swedish varves from North and

South America, the Himalayas, East Africa and Ireland – but few are prepared to accept his evidence as conclusive.

Perhaps the strongest argument against the value of de Geer's teleconnections comes from the man who, in a sense, had most to gain by believing in them. The great North American exponent and practitioner of varve-counting has been E. Antevs. By varve-counting he studied the retreat of the American ice sheets from Long Island north right up into Ontario. Working throughout the late 1920s and early 1930s he did more than any man to provide a chronology for the last 15,000 years on the North American continent. His conclusions have been disputed in many respects by workers in the same field – and he has eventually been shown to have been wrong by the archaeological clocks when they were discovered. But like de Geer in Sweden, Antevs provided something that had never been provided before, a chronology which people could at least argue about, even though he could not provide a link between his varve-series and the present day, even though there were gaps in his various series where estimates had to be made.

Yet it was Antevs himself who denied the possibility of any valid connection between the North American varves and the Swedish patterns, at least as far as the available evidence went at the time (1935). Quite apart from the lack of evidence in the varve-patterns, there is the theoretical argument that there is no justification for expecting the pattern of summer warmth in North America to bear any recognizable relation to the pattern of climate in Scandinavia. At the time when the first universal clock was, in fact, just beginning to take shape in a laboratory in the American Midwest, Zeuner rather gloomily summed up the varve chronology story:

It is necessary however to emphasise that the accuracy expressed by the use of dates A.D. and B.C. is largely fictitious. Whether the Baltic Ice Lake was drained in −1073 = 7912 B.C. is doubtful but it is most convenient to accept some such date to construct the time scale on. De Geer himself has frequently used round figures instead of accurate dates . . . Varve chronology promises to produce results and the time scale for the last 10,000 years appears to be trustworthy. But more systematic re search is needed to strengthen and to straighten out the fabric of cross-dated local chronologies.[9]

This was not, however, the sum total of the contribution made by varve chronology. Whatever its internal inconsistencies and inaccuracies the whole structure seemed 'trustworthy' and the varves could be related to several major geological trends. Such events as the draining of the Baltic Ice Lake left traces in other ways as well as varves. As the ice slowly melted the depression in the land which is now filled by the Baltic Sea began to fill with water, although the ice sheet still spread well to the south of this half-frozen lake on the ice. Eventually, however, the ice retreated far enough to let this vast amount of pond water go rushing out into what is now the North Sea, though the outpouring did not take place through what is now the mouth of the Baltic between Denmark and Sweden, but went right across what is now the land-mass of southern Sweden, through the so-called Billingen Gap. Naturally such a catastrophic event has left its traces in the geological record, and the entry of salt water into the Baltic depression shortly afterwards has left even clearer traces in the preserved shells of saltwater molluscs appearing for the first time. The retreat of the ice sheet left further marks, first in the rise of sea-level as the vast amount of water locked up in the ice sheets poured into the oceans, and then in the comparative fall of sea-levels as the land rose when the weight of the ice shifted from it. These changes are most clearly marked by successions of former seashores now clearly distinguishable at many different heights above the present sea-level; these are called 'raised beaches'. Raised beaches can be related to series of varves, and on raised beaches the camps and rubbish heaps of prehistoric men can be found. Often the vast quantities of oyster and mussel shells show clearly that the men who lived on these now raised beaches must have been living on a beach that was, in their time, at sea-level. From the link with the varves a time can thus be calculated at which the camps were occupied.

Calculating a date by relating a raised beach to a set of varves could never provide a very accurate result, nor could the result be applied to any much larger area. The next timing-device to be discovered – the technique of pollen analysis – offered hope of being applicable over wider areas. It was never suggested very seriously that pollen analysis would provide an archaeological 'clock' of great accuracy, unless by some lucky chance pollen deposits could be tied in with the varve-series. But as an entirely new method of providing a relative dating technique, a technique which was entirely 'natural' and did not depend upon manmade artefacts, it was at first hailed with great enthusiasm. (Pollen analysis has continued to be of enormous importance in modern archaeological studies for the wide variety of information it can give about prehistoric environments, though it is little used for 'timing' purposes nowadays.)

Pollen analysis was another Scandinavian discovery, and, like several major advances in archaeological thought, it sprang from the remarkable preservative qualities of the Danish peat bogs. In our own recent times the discoveries of Tollund Man and Grauballe Man – wonderfully preserved human bodies more than two thousands years old complete with clothes and analysable stomach contents – have aroused considerable interest, and have done great work in popularizing archaeological discovery. But there have been similar discoveries of well preserved bodies in the past, and the one that aroused professional interest in the early days of archaeology was a well-recorded finding of a woman's body in Haraldskjaer Fen near a former Danish royal palace at Jelling, in 1835. The local antiquarians – for this was before archaeology proper had been born – promptly decided that the lady must have been Queen Gunhild, the Norwegian widow of King Eric Bloodaxe, who had been invited to Denmark in the tenth century to marry King Harald. It was a matter of known history that, on her arrival, she had been thrown into a bog instead of a marriage-bed. We have no reason to suppose now that the body was that of Queen Gunhild, but the details of the find were well known and a Danish Almanac of 1837 explained, 'There is a strange power in bogwater which prevents decay. Bodies have been found which must have lain in bogs for more than a thousand years, but which, though admittedly somewhat shrunken and brown, are in other respects unchanged.'

It was a Swede, Lennart van Post, who followed up these clues and started examining the comparable bogs in southern Sweden in the early years of the twentieth century. He soon discovered that the peat bogs preserved, not only human bodies, but pollen grains from different plants and trees. Since the pollens of different plants differ markedly in size, shape and structure, he could identify the different types even in layers of peat which, by reason of their depth below the surface, must have been laid down thousands of years before the present. Furthermore, van Post spotted the fact that the proportions of different types of pollen varied between different layers of peat, and often varied directly in proportion to the depth at which they lay.

> Pollen grains are the male reproductive structures produced by seed bearing plants. Their designed function is to transport the male gametes or sperm to the female egg where fertilization can occur. For each pollen grain that achieves this function thousands fail and are instead deposited on the earth as a pollen rain. Once deposited, many grains are oxidised or destroyed by microorganisms, certain types of fungi, and other forms of mechanical and/or chemical weathering. Yet some earth environments are suitable for pollen preservation and here the wall of the pollen-grain survives and becomes part of the fossil record.

This admirably precise and clear summary comes from a paper by Vaughn M. Bryant and Robert K. Holtz of the University of Texas.

Van Post's conclusion was that an analysis of the proportions of different types of pollen at any one level of his peat bog would give an accurate picture of the types of plants growing in the neighbourhood at the time the peat layer was formed. If he analysed the pollen at various different levels he would get a picture of the way the various pollen-producing plants had increased or decreased in numbers as time passed in that locality. His expectations proved justified. Many others took up the technique and eventually it was possible to see quite clearly how, as the ice had retreated, the area was colonized first by the plants we associate with arctic tundra conditions, then by dwarf birch and pine, until finally the great trees of European woodlands appeared, elm and beech and oak, and bushes such as hazel. As the work with the new technique went on, several crucially important peat bogs were found where the bottom layer rested on a raised beach or even on a bed of clay containing telltale varves. Thus the pollen record of the changes in vegetation in Scandinavia could be linked with features which fitted into a known time-scale and even approximately dated. And since the plant-life of an area reflects the climate, a picture of the climatic changes in north-western Europe since the end of the Ice Age was built up.

Even without dating by cross-reference to varves, the pollen analysis therefore allowed a rough chronological framework of differing climatic periods to be built up and this was found to apply to North America as well as to Europe. Clearly defined phases of climate could be seen such as the Boreal and Atlantic, the Sub-Boreal and Sub-Atlantic; Boreal being warm and dry, Atlantic being warm and moist. Each phase had a distinct vegetation pattern especially in the forested areas; thus the warm, dry Sub-Boreal was marked in the north-east of the U.S.A. by a predominance of oak and hickory trees; while the cool, wet Sub-Atlantic phase which followed it had a broader vegetation of oak, chestnut and spruce. Plainly, these changes in vegetation and climate must have affected the modes of life and the food-gathering activities of the men who lived through them. We guess that prehistoric communities might well have moved with their traditional food sources either northwards or southwards, rather than have remained in one area and changed their habits and traditions. In fact pollen analysis did little towards actually dating the archaeological sites at which it was carried out – but it had the far wider and more important effect of making the archaeologist realize that the environment in which men lived was every bit as important to his study of prehistory as the tools that prehistoric man used.

This, however, was a slower, longer and later development.

Pollen analysis, seemed, in the years immediately preceding the Second World War to offer some positive hope of providing a rough timing system, at least in western Europe. It had become clear that the northwards spread of the beech tree could be clearly followed as it emerged from its strongholds in places like southern Germany and advanced in the wake of the retreating ice. Similar hopes were also placed on tracing the advance of the elm tree which presented very sharp 'peaks of prominence' in the diagrams of pollen-abundance, and which seemed to be more closely linked with man's activities than the beech. But both hopes were to prove vain, and it is nowadays held that the spread, and later decline, of the elm is more likely to have been caused by man's activities than to have affected man.

Pollen analysis failed to provide a clock mechanism, but has instead become one of the most important tools of the modern archaeologist for entirely different purposes. The way was shown in the closing years of the 1930s by the Dane, Professor Johs. Iverson. He was one of the first to use pollen analysis as an archaeological tool rather than as a substitute clock. He noticed that his pollen records from various Danish archaeological sites showed an unexpected change in their nature at about the time when some change should in every case be expected as the Atlantic climatic period gave way to the warm and dry Sub-Boreal period – a time which he estimated to be about 2500 B.C. He saw that it was not so much the varieties of tree pollen that changed as the sudden decline in the proportion of pollen from trees. At the same time the pollen of lowlier plants and grasses suddenly increased in proportion. He was soon able to show that among the increasing pollen of herb plants (as opposed to trees) there was the pollen of cultivated plants. Agriculture had therefore been brought to Denmark at that time, and since the sites from which the pollen came were Neolithic sites, it was Neolithic man who brought agriculture to Denmark. Further study of his sites and his pollen analyses showed that the decline in tree pollen was closely associated with a layer which contained a remarkably high proportion of charcoal, and that immediately above this layer of charcoal the pollens of cereal plants, cultivated cereal plants, began to appear. Neolithic man, it was clear, had burned the forests – presumably in the slash and burn technique used by some primitive tribes to this day – and had planted his first crops in Denmark in the land he had so cleared.

Iverson's important results in this field were published in 1941 and like many another scholarly piece of work were lost for several years in the confusion of the Second World War. It was not until the start of the 1950s that his application of pollen analyses to archaeological sites for purposes other than dating were seriously followed in the U.S.A. and Britain.

So it was clear as archaeology and prehistoric studies began to start again at the end of the Second World War that no satisfactory archaeological clock had yet been found, although the best estimates of the various semi-satisfactory systems could be combined, as Zeuner had done, into a broad chronological framework which at least gave a consistent system of guidance as to the order in which the major developments had occurred and some idea of the length of time taken by these developments. There had been other methods suggested than those considered in detail here – estimates based on the length of time required for the evolution of animal species, and attempts to trace the effects of astronomical features such as the sunspot cycle of 11.4 years or the precession of the equinoxes, in the geological record. But these had never offered much hope for dating in man's prehistory and had been of only marginal value in helping to sort out the chronological basis of 'geological' time, the time-scale of the evolution of life and the development of the planet we call Earth.

Nevertheless, there were interesting presages of the future even when Zeuner wrote his first edition, in 1946, when Libby was just starting on the research which would lead to the first practical archaeological clock. It was, in fact, already clear that somewhere in the process of radioactivity there were clocks to be found. Radioactivity, the natural and spontaneous disintegration of certain atoms, in which the atom changes from one element into another while emitting either an electrically charged particle or a burst of electro-magnetic energy, had been known since long before the start of the Second World War. The investigation of the phenomenon and the attempt to discover what actually happened had occupied a fairly large number of the leading physicists and chemists of European laboratories in the late 1930s and had also been taken up in the U.S.A. The best-known process was that by which radium atoms eventually became lead atoms – indeed the study of this process had begun as far back as Marie Curie's famous discovery of radium in 1898. The long series of disintegrations by which a uranium atom eventually turns into a lead atom was also known perfectly well and had been openly published in scientific literature. It was known, too, how slowly these processes took place in nature; it had been measured that one gramme of radium would take 1590 years before half of its mass had decayed. Other decay rates had been calculated in hundreds of thousands and millions of years.

Some of the radioactive disintegrations give off their radioactivity in the shape of 'alpha-rays', which are particles of the same nature as the nucleus of an atom of helium with its electrons stripped away. If such a radioactive disintegration occurs in the middle of a rock, the alpha-particle will soon strike other atoms, will pick up spare electrons, and will become an atom of helium. A rock specimen which contains radioactive atoms will thus come to contain, over millions of years, a measurable amount of helium gas. In a rock which contained uranium at the time of its formation there will, therefore, in modern times, be less uranium than there was originally; there will be an accumulation of lead, derived from the uranium by disintegration, and there will be helium produced by the alpha-particles. Hence the proportion of helium to remaining uranium, and the proportion of lead to uranium must bear a definite relation to the age of the rock.

This had been worked out before the Second World War, and there is no mention of wartime nuclear work in Zeuner's review of the method. There were, of course, snags – the amount of natural lead originally present at the formation of the rock would complicate the calculations; there are other elements which can produce lead by radioactive disintegration such as thorium; and the helium, being a gas, can work its way through the crystals of the rock and escape to the atmosphere. But careful selection of samples, by choosing rocks which contain almost no natural lead, and sophisticated mathematical techniques, can overcome these problems, and, as long as the rates of disintegration are known, an answer can be found for the age of the rock. In the late 1930s a small number of quite satisfactory results had been obtained using these methods.

The radioactivity methods were highly promising for calculating geological ages of rocks. But to calculate that a rock is 1,250 million years old is hardly 'dating' in the sense that the word has been used so far. And the radioactivity of rocks gave no sign, in 1946, of helping to 'date' an archaeological site that necessarily fell in man's short span of one million years.

So, only twenty-five years ago, with one small exception, there was no archaeological clock in existence. It was not just that the many methods suggested for dating archaeological material had turned out to have weaknesses or to be so inaccurate as to be useless in practice. It was also the case that all methods without exception could be applied only locally even if further research could perfect them – the drain-off of water from retreating ice sheets in Scandinavia could not possibly relate to the activities of Egyptian pharaohs, early Chinese farmers, or the pre-Columbian inhabitants of Peru. Nothing that looked promising as a universal archaeological clock could be seen anywhere.

1. G. Bibby, *The Testimony of the Spade*, p. 33.
2. L. Woolley, *Digging up the Past*, p. 26.
3. ibid., p. 27.
4. C. Coon, *Seven Caves*, p. 141.
5. ibid., p. 142.
6. G. Daniel, *The Megalith Builders of Western Europe*, pp. 70–75.
7. F. Zeuner, *Dating the Past*, p. 1.
8. ibid., p. 19.
9. ibid., p. 45.

3

Tree-rings

The most unusual feature of the uneven history of tree-ring dating – officially entitled dendrochronology – is that the precise date on which this technique emerged as a useful archaeological 'clock' is known – the evening of 22 June 1929. This precision is the more unusual in that the man who established the new technique started the work with no intention whatsoever of doing anything of the sort.

It has been known for many hundreds of years that trees produce one ring for each year of their growth. When a tree is cut down anyone can see that the structure of the wood consists of a series of ever-increasing circles or dark and light rings, starting from the very centre and working steadily outwards to the bark. In the third century B.C. Theophrastus wrote about this and speculated about the cause. We now know that at the start of each growing season, in the spring, the tree manufactures layers of large cells underneath the bark – the material is technically known as cambium. It continues doing this throughout the time of active growth, but as the summer goes on the cells that are manufactured become smaller and smaller. In the winter growth stops, even for evergreen trees like pines, and the cells laid down during the preceding year die and become wood. The following spring the same process is started again and so the concentric circles or 'tree-rings' are formed, with the pale parts formed of the large cells of springtime and the dark portions made by the smaller, more densely packed cells of late summer. By counting the rings of a freshly cut tree across the stump of the trunk it is within anyone's power to calculate when the tree started its life – seventy rings on a tree cut down in 1972 means that it was planted and started life in 1902.

But a closer examination of the rings of a tree will soon show that they vary in thickness, and it is immediately clear that this can tell us about the climate of the years through which the tree has lived. A thick ring means that in that particular year there was much growth, and that means, for most trees, a summer when there was a sufficiency of both water and sunshine. A thin ring means there was little growth in that year and the most likely reason for the lack of growth would be lack of water – a year of drought probably, though it could also mean a cold and sunless season.

It has been noted that the first suggestion that archaeological sites might be dated by counting the tree-rings was made by De Witt Clinton in New York in 1811. But nothing was ever developed from this and it was the variation in the size of the rings carrying information about the variation of climate that first attracted the attention of Andrew Ellicott Douglass, the man who is universally accepted as the founder of the science of dendrochronology. Douglass was a man in the great tradition of American pioneers. He was born in a Vermont parsonage in 1867, and his middle name Ellicott was given him in honour of his paternal great-grandfather who had helped in the laying out of the new capital city of Washington. His father and both his grandfathers had been college presidents, and the young Andrew Douglass started out brilliantly in an academic career. He chose to become an astronomer, though he had graduated with honours in physics and geology as well as astronomy, and he was appointed to an assistantship at Harvard College Observatory in 1889.

In 1894 Percival Lowell asked him to select a site for the Lowell Observatory under the clear and brilliant skies of Arizona, and it was Douglass who selected Flagstaff, Arizona, as the site of the now famous Observatory and became the first assistant there. This was pioneering work – travelling in a buckboard in wild terrain in an area which had not then even achieved statehood – it was still the Territory of Arizona.

Douglass's professional astronomical interest at the time was in sunspots and in the sunspot cycle of 11.4 years. In order to establish that this cycle existed, and had existed in past times, Douglass needed some record of sunspot activity in the past. It is possible that it was the contrast between the greenery of his native New England and the arid, but not desert, conditions of Arizona that turned Douglass's thoughts towards the possibility that tree-rings might contain some record of past climatic changes. He hoped, of course, to find some pattern in the variation of thickness of the rings that showed a regular cycle of eleven or twelve years, to confirm that the sunspot cycle had existed in the past. Starting in 1901, Douglass soon had a curious mixture of tree-ring specimens and telescopes in the Flagstaff Observatory.

He travelled considerably outside the Observatory to collect his specimens, going especially to places where lumbering was going on, and making rubbings or tracings of the freshly cut stumps. He soon became aware of certain regularities in the patterning of thick and thin rings which turned up again and again in Arizona timber – and once in 1904 he recognized one of these patterns in the outside rings of an old tree stump and was able to announce with great confidence the exact year in which that tree had been felled, which, not surprisingly, astounded the farmer who had been responsible himself for the felling of that tree. By 1907 Douglass was able to make the first of his many publications on tree-rings, presenting a long and firmly established pattern of variations which showed in the trees grown around Flagstaff.

A couple of years later, in 1911, when he was in Prescott, Arizona, fifty miles south-west of Flagstaff, he recognized in the local timber patterns he had memorized from his Flagstaff trees. It was then that the possibility of using tree-rings for cross-dating from one place to another dawned on Douglass and he entirely reoriented his research in this new direction. He had moved to the University of Arizona in Tucson some years previously and had become Professor of Astronomy and Physics, and in this particular year of 1911 he was Acting President of the University. He started collecting samples from groups of trees in Europe to see if he could find any similarities between them and North American tree-rings. More importantly he started studying the giant sequoias of California, which were known to be the largest organisms on earth and probably several thousand years old in some cases. Hopes that the sequoias would show links with the trees of Arizona proved vain by 1919, but, before then, another and more important sequence of events had started.

In 1914 Douglass had described the possibilities of cross-dating and counting tree-rings at a meeting in Washington. Among the audience was Clark Wissler of the American Museum of Natural History and, at the end of the First World War, Wissler got Douglass to look at some wood specimens from a prehistoric ruin in New Mexico. Douglass asked for more and the next year, in 1919, he received six further beams from the so-called 'Aztec Ruin' in New Mexico. He was able to announce almost immediately that there were distinct cross-datings to be found in the tree-ring patterns of the timbers from this ruin. In 1920 Douglass officially joined in the investigation of the Pueblo Bonito ruin, which is one of the most spectacular remains left by the American Indians of the south-west U.S.A. This particular investigation is important in the history of archaeology as the best early example of an interdisciplinary examination of a site – a model for what was to come later. But at that time it marked simply the start of an extensive process of cross-checking the tree-rings that could be found in the beams and timberwork of a series of nearly forty prehistoric sites of the 'Pueblo' culture. In the next few years Douglass was able to establish the chronological sequence among these ruins and to build up a chronology, a year-by-year system of markers covering nearly one thousand years. The big problem that remained was to establish some relation between this chronology, floating anchorless in the stream of time, and modern dates. When exactly in terms of years A.D. and B.C. did this set of one thousand or so years belong?

The Pueblo sites were, of course, 'prehistoric', but this word means no more than 'before A.D. 1540' in the south-western U.S.A. History begins when written records, available to us now, begin.

For north-western Europe, history begins just before the start of our era, when Julius Caesar wrote of his campaigns in Gaul and Britain. History on the west coast of America begins in 1540 when the Spanish missionaries arrive. But there are two further facts about the south-west of the U.S.A. which are important in this context. First there is the climate, warm and dry, which means that wood is preserved and remains usable for many years, far longer than in the wetness of north-east America or western Europe. The climate also provides that the majority of the trees in the American South are pines – *Pinus ponderosa* or yellow pine, and thus few difficulties arise comparable to trying to compare the growth rings of oak with those of ash or elm. The other relevant factor are the Indians themselves – the Hopi. They are a very pacific people who have developed a culture extremely well suited to their environment, a culture which has changed remarkably little even under the impact of white invaders. Present-day Hopi live lives very similar to those of their ancestors of a thousand years ago. Perhaps because of their peaceableness they never got engaged in war with the white Americans and so they have survived where many more belligerent tribes have virtually disappeared. Indeed the only major event in the history of the Hopi in the last five hundred years, apart from the arrival first of the Spaniards and later of the United States, was their uprising in 1680 when they burnt the Spanish missions.

There is, indeed, a village near the Little Colorado River, called Oraibi, where the Hopi live now and where they are believed to have lived since before the arrival of the Spaniards in 1540. Some of the logs that support the roofs of their typical Pueblo houses were clearly cut by stone axes and so have been preserved and in use since before the first metal tools arrived. (Douglass found a ladder in which one pole had been cut about 1570 and the other pole in 1720 – so something had got broken two hundred and fifty years ago.) It was to Oraibi that Douglass and his assistants went in their search for cross-datings which would anchor their 'floating' prehistoric chronology to our present count of years. Taking samples from the village timbers, which plainly included beams taken from the Spanish missions which the Hopis had salvaged and turned to domestic use after the affair of 1680, Douglass was able to build up a continuous record of tree-ring patterns, cross checking one beam against another until he got back to somewhere about the year A.D. 1400. Then he began searching the village for really old pieces of wood to push the series even further back. The oldest piece he found was in part of the village which the Indians had abandoned in 1906. It was still supporting the roof of a half-ruined house, standing as a central pillar. It could not, therefore, be cut, and the new technique of taking a cylindrical drilling right through the middle had to be adopted. It turned out that this log must have come from a tree which was felled in about 1370 and it took the main chronology back to 1260.

At this point progress became much slower, and nothing seemed to push the established chronology, stretching from modern times, back any further. They went to other villages and ruins. Nothing seemed very useful, except some finds of big lumps of charcoal in the ruined village of Kawaiku, which confirmed and strengthened the master-plot of tree-ring patterns between 1300 and 1495. These master-plots were embodied in extremely long strips of paper with a continuous line tracing the varying thicknesses of the rings from year to year in graphical form. It was therefore comparatively simple to see when a series of peaks and hollows on one strip of paper exactly matched a similar series of peaks and hollows on another strip. Douglass had established in many samples that there were important markers in the system, such as six very thin rings which occurred in the 1280s and 1290s when there must have been terrible droughts.

It was not until the early summer of 1929 that Douglass started work in the village of Showlow, among the Indian houses. There he found a beam that was very large and seemed very old, although the end had been burnt off in a cone-shape. It was soon clear that the outer rings gave a very clear trace of the correct patterns for the years 1300 to 1380. Working inwards there were the unmistakable six drought years of the end of the thirteenth century, and then, to their delight, the investigators found that, working back towards the centre, they reached 1237 – the earliest yet.

But it was not until the evening, when they were looking at the master records of their current investigation and of the prehistoric series from the Pueblo culture, that Douglass suddenly saw where the correspondence occurred. This was the famous evening of 22 June 1929, and what he realized was that the ring on the beam they had sampled that day in the village of Showlow which they had now calculated to be from 1251 matched exactly the 551st ring on their prehistoric strip. They had, in fact, long since bridged the gap between their 'modern' chronology and their floating 'prehistoric' chronology, but none of them had spotted it. And now the south-western states of the U.S.A. had a chronology in years stretching back to A.D. 700 and the forty ruins of the Pueblo culture they had earlier investigated could now be assigned precise dates in years A.D.

Work in the south-western states done by the Department of Dendrochronology of the University of Arizona and others has pushed back the tree-ring dating of archaeological sites in that area to around the beginning of our era and dates can now be given to the culture that preceded the Pueblo, called the Basket Maker culture.

But the provision of a chronology for the south-western United States was the last major triumph of tree-ring dating for nearly forty years. It proved impossible to relate the patterns of the rings in New Mexico or Arizona timber to other sets of rings. A clock had indeed been found, but it was a very local clock, applicable only in the area where it had been found. Teleconnections could not be made either. There were claims that wood from a Swedish fort in Gotland compared with the Californian sequoia tree-ring pattern and dated the fort to the fifth and sixth centuries A.D., but this work is not widely accepted and has not proved possible to repeat elsewhere.

However, tree-ring dating has been carried out all over the world – in Japan and Russia, in Egypt, Turkey, Germany, Britain and Denmark as well as in many other parts of the U.S.A. and

Alaska. It is being found more and more useful as time goes on – but only in certain suitable places. In Alaska, where wood tends to be well preserved, tree-ring dating has been highly successful, and this is one of the few places where 'absolute' chronologies have been set up – that is lists of years stretching back from the present, enabling a single piece of wood to be assigned to actual dates in years.

But 'floating' chronologies based on tree-rings have proved much more useful in general than absolute chronologies. These floating chronologies may, for instance, help to sort out the order and method of construction of Bronze Age forts in Germany or Neolithic sites in Switzerland. They are even more useful when applied to research into the development of medieval towns in Europe where so much of the building was done in wood. In these circumstances a short, floating chronology which can be linked into known historic events in the life of the town can be used to date the construction of particular buildings accurately to within a very few years.

In temperate climates, however, like that of Britain, where the weather can vary so much, it is often possible to show definitely that there is no correlation between the tree-ring patterns of trees growing within a very few miles of each other. Furthermore, there are many more species of trees to be brought into consideration than in Arizona, and this sort of problem does not seem strange to those of us who live in countries where the weather can differ from one side of a hill to the other. At the opposite extreme are the countries where timber is very scarce or non-existent, and where, therefore, wood is hardly used as a building material. Unfortunately for tree-ring dating this applies to much of the classical area of archaeology, the Middle East. Apart from the cedars of Lebanon – indeed this is the reason why they were so valuable and so much extolled – Syria, Iraq, Palestine, Egypt and much of Persia are virtually treeless and seem to have been so since before Neolithic times. Mudbrick is the building material of ancient Mesopotamia, later replaced by the more durable fired brick. Stone, as we know so well, is the material of the great Egyptian monuments, pyramids and temples.

So dendrochronology was not the answer to the search for a universal 'clock that runs backwards'. It could only provide a clock mechanism with great labour in certain selected places. But in those places where it can be made to work it is highly accurate and reliable. Its greatest long-term value came from those virtues when the clock of south-western U.S.A. was used to 'set' the timing of the really universal clock. But that is part of the story of radiocarbon dating, and the triumph of the tree-rings comes in Chapter Five.

4

The Clock Discovered

The first universal clock of archaeology, the first 'clock that runs backwards', was radiocarbon dating. It is still the most important and valuable clock-mechanism available to history and archaeology. It has been described by Glyn Daniel as 'perhaps the greatest breakthrough in the development of archaeology'. It was discovered in 1949, by the American, Willard F. Libby. It measures time by using any material that was once part of a living creature and it achieves universality because it uses the natural processes of life and death as the principle of its operation.

It is often said and written that radiocarbon dating rose out of the wartime work on the atomic bomb. This is not true. It is a misunderstanding of the nature of physics and of the history of that science. Radiocarbon dating is a product of the mainstream of the science of physics; its roots are clearly discernible in the knowledge gained in civilian, university physics laboratories before the outbreak of the Second World War; the basic discoveries that allowed the concept of radiocarbon dating to be formed had been made before 1940; no work on the technique was done in wartime weapons research. The only benefit radiocarbon dating has received from atomic weapons research has been in the development of instruments and methods of handling radioactive substances, and in the development of isotope-chemistry in wartime laboratories.

The mushroom-shaped clouds from nuclear weapons distort our view of the physics of the 1920s and 1930s. To us the most important political fact of our times is that mankind possesses the knowledge and technology to destroy itself, and this knowledge sprang from the work of the physicists in the 1920s and 1930s, and was converted by physicists into the technology of total destruction between 1940 and 1945. Yet it is also permissible to view the history of the science of physics in our century as one of the most exciting intellectual processes the world has ever known. Physics has changed man's view of his world; it has revolutionized our ideas of what makes matter, of the structure of ourselves and all that we see and hear and touch. The atomic bomb is a ghastly by-product of this intellectual ferment, its development was but a digression in the process that was changing the concept of the universe just as surely as Galileo and Newton changed it three centuries ago. It was the intellectual excitement, and not the prospect of the atomic bomb, that attracted the physicists to atomic studies in the 1920s and 1930s, led by some of the greatest intellects of our time – Einstein and Lawrence in the U.S.A., Rutherford and his team of brilliant young men such as Chadwick and Cockroft in the Cavendish Laboratory in Cambridge, the devoted, courageous Curies, husband and wife, in France, Hahn, Strassmann, Meitner and Heisenberg in Germany, Flerov and Kurchatov in Russia, Niels Bohr in Copenhagen, with the Italian, Fermi, livening it all up with a series of brilliant, unorthodox ideas.

The chemists had shown that all matter was made up of atoms of ninety-two different elements. This was one of the great intellectual achievements of the nineteenth century. It had further been shown that the atoms of the ninety-two elements could be arranged in order of increasing weight, starting from hydrogen, the lightest and working up to uranium, the heaviest.

The essence of the physicists' achievement is that they showed that all the different types of atoms had essentially the same sort of structure. In the centre there is a heavy body consisting of two sorts of particles – protons each with a positive electric charge, and neutrons with no charge at all – otherwise identical. Round this heavy body electrons circle in orbits, each electron very light in weight but with a negative electric charge. The number of negative electrons in an atom exactly equals the number of positive protons in the 'nucleus' of the atom. One element, say sodium, differs in its chemical behaviour from another, say chlorine, only because it has a different number of electrons in orbit round the nucleus and because it has a different weight caused by the different number of protons and neutrons in the nucleus.

There was one important qualification to this essentially simple system. This was the existence of isotopes, atoms which have a different number of neutrons in their nucleus from the normal type of atom of that particular element. The extra, or fewer, neutrons make no difference to the chemical properties of the atom – it is still chlorine or sodium because the number and arrangement of electrons is still the same and the number of protons in the nucleus is also the same. But the weight of an atom of an isotope is different from the weight of the normal atom of that element, and the isotope can, in fact, be separated from the normal element by using the mass-spectrometer, a device which is very sensitive to different weights in individual atoms.

At the simplest end of the table of elements normal hydrogen has a nucleus of one proton, and one electron orbits around it. If the proton in the nucleus is joined by a neutron, there is still only one electron and the atom is still hydrogen – but it is now 'heavy' hydrogen. It joins with oxygen in the normal way of two hydrogen atoms to one oxygen atom, but the heavy hydrogen makes it 'heavy' water. A further neutron can also be added to the nucleus of hydrogen to give hydrogen 3, which is also called 'tritium', and is the stuff of 'hydrogen bombs'. Helium, the second element in the table, is normally helium 4 with a nucleus of two protons and two neutrons (hence the 4) and with two electrons in orbit. One neutron can however be taken away to leave the isotope helium 3 – still helium because it has two protons and two electrons.

The 'normal' atoms which make up most of our world are, in fact, only the most stable versions of each type of atom. Radioactive isotopes are rare in nature because they are, by definition, unstable. When they disintegrate they emit particles such as neutrons or electrons, or they give off energy and turn into other sorts of atoms.

Some radioactive isotopes are so unstable that they disintegrate in fractions of a second, others may remain stable for millions of years. But the energy or particles given off by any atom disintegrating or turning into another sort of atom is 'radioactivity'. The discovery of radioactive isotopes and their instability therefore tied in well with the parallel results in the discoveries of radioactivity initiated by the Curies stemming from their finding of radioactivity emitted by the element radium, whose atoms steadily turn into lead atoms.

By the late 1930s all the natural elements had been discovered and isolated and many of their less common isotopes had also been discovered. But many more isotopes could theoretically exist on the basis of the structure of the atom which had been postulated. The next step was obviously to bombard 'normal' atoms with particles such as neutrons and see what new forms of matter could be created. An interesting target would be the heaviest atom of all, uranium 238. (Normal uranium is uranium 238 which is quite stable. It is the rare radioactive isotope uranium 235 – uranium minus three neutrons – which is responsible for most of the natural radioactivity of uranium and which is most likely to 'fission' when bombarded with neutrons.) It was an entirely 'pure science' experiment in bombarding uranium with neutrons which led to quite unexpected results which were later seen to point the way to nuclear weapons.

These broad studies of the nature of the atom, isotopes and radioactivity did not, of themselves, lead to the dicovery of radiocarbon dating. But they provided the vitally necessary background of understanding of the nature of matter which enabled the results of other experiments to be interpreted in such a fashion that the possibility of radiocarbon dating sprang to one man's mind.

These other experiments, the direct origins of radiocarbon dating, were investigations of the upper atmosphere of the earth, and in particular attempts to measure the effects of the impact of cosmic rays on the earth. Libby himself traces his work back to this origin. 'Radiocarbon-dating had its origin in a study of the possible effects that cosmic rays might have on the Earth and the

Earth's atmosphere,' he said in his official Nobel Lecture on 12 December 1960 as he received science's highest award for his discovery of the first 'clock that runs backwards'.

The success of Marconi's first transatlantic radio broadcasts, which were theoretically impossible according to the knowledge of the time, had revealed the existence of the ionosphere, the electrically charged layers of the upper atmosphere which bounce radio signals back to earth. Twenty years of research had shown that the ionosphere received its electric charge from streams of incoming charged particles most of which had their origin in the sun. But a few of the particles that bombard the earth are cosmic rays – particles of colossal energies coming from unknown places far away in the universe. We still do not know very much about cosmic rays, and we still do not know what exactly happens when they first strike the outermost, thinnest layers of our atmosphere. It is reasonably clear that the ionosphere acts, not only as a mirror for radio waves sent up from the surface of the earth, but also as a shield protecting the tender life on the surface of the earth from the impact of all these bombardments. Down on the surface of the earth all that we can observe of the impact of cosmic rays is the occasional shower of charged particles and neutrons which are the second-order or third-order results of very energetic processes going on far above us.

It was in trying to find out more about these phenomena that Professor Serge Korff of New York University produced the first link in the chain of argument which led first to the theory of the radiocarbon clock and finally to its application. He sent up neutron counters in a balloon (neutron counters being simple devices which recorded the arrival of neutrons and their approximate energies). This showed that at heights of sixteen or so miles above the earth the air was full of neutrons speeding down with energies of about a million electron volts. Whatever actually happens when a cosmic ray strikes the outer atmosphere of the earth with an energy of billions of electron volts, one of the principal results is a shower of less energetic neutrons. There were, at the first count, about two of these neutrons coming down through every square centimetre each second. Since we, living on the earth below, are not bombarded with neutrons at this rate, the neutrons must react in some way with the atoms of the air which, of course, gets denser and denser, and more likely to stop the neutrons, the further down one gets towards the surface of the earth.

No laboratory on earth could then – or can now – produce the equivalent of cosmic rays at billion-volt energies, which is the chief reason why we know so little about what happens at the top of our atmosphere when the cosmic rays strike. But it is comparatively easy to get million-electron volt neutrons, and laboratory studies on the effect of such neutrons in air had already been performed in other centres when Korff found the high-altitude neutrons that were the secondary effects of cosmic rays. So it was known in 1939 that the effect of neutrons in air is mainly to produce radiocarbon, that is the isotope carbon 14. About one per cent of neutrons in air form radioactive hydrogen 3, tritium, and there is also a minute production of helium 3; the remainder of the neutrons form atoms of carbon 14.

The oxygen in our atmosphere is more or less inert to bombardment by neutrons but the commonest gas in the air is nitrogen and that is usually in the form of nitrogen 14. A neutron hits the nitrogen nucleus and lodges in the nucleus itself, knocking out a proton which takes with it one of the negatively charged electrons. This leaves an atom which is in fact a carbon atom with an excess of neutrons in its nucleus. This is the carbon 14 atom, and because it has the excess of neutrons in its nucleus it is unstable compared with the normal carbon atom which is carbon 12. (There are also carbon 13 atoms which are stable and which occur in the proportion of one to every 99 carbon 12 atoms.) The carbon 14 atom disintegrates – or at least its nucleus disintegrates – in a process which can best be described as one of its neutrons turning into a proton and emitting an electron – a beta-ray. This turns it back into an atom of nitrogen 14.

All this was known to the world of physics at large in 1939 and, in one of his papers announcing the discovery of the neutrons at high altitude, Professor Korff pointed out that the principal way in which the neutrons would disappear before reaching the earth's surface was by the formation of radio carbon. But in 1939 no one had ever spotted radiocarbon in living matter; it was quite unknown in nature and radiocarbon had never been detected except in these laboratory experiments studying the impact of neutrons on air.

In any case this was in 1939 and the impact of the Second World War was already making such esoteric laboratory studies as nuclear physics and research into cosmic rays seem something like escapism – many of the leading British physicists were already being moved into the radar field, for instance.

But something about Korff's work must have stuck in the mind of a young chemist in California – and this is where Willard Frank Libby comes into the story. Although he is normally friendly and helpful, Libby is a rather austere man, very much the public's idea of a scientist, tall and thinnish, nowadays balding, with a great rising forehead and dark-rimmed glasses. Not a man given to recounting anecdotes of how he suddenly had his great idea in the bath or on the golf course; instead a scientist who, when asked to speak or write on his subject, says what he has to say briefly and rather baldly, but clearly and carefully. The idea that radiocarbon dating might be a possibility seems to have come shortly after he read of Korff's work and slowly 'gelled' (his word) into a coherent theory during the years 1941 to 1945, the war years as far as American scientists were concerned. He could do nothing about his theory until the war was over and he could return to university life and 'civilian' research.

Willard Libby was one of the younger generation of scientists in the 1930s and he was a chemist rather than a physicist. He was born on 17 December 1908 in Grand Valley, Colorado, and graduated as a Bachelor of Science (B.S.) at the University of California in 1931. He went into research at the University of California and obtained his Ph.D. in 1933. He plunged straight into the booming field of radioactivity – his first published work was about amplifiers for Geiger counters (for measuring radio-activity) and his third paper put him right in the forefront of the new developments in 1933; it was called 'The Radioactivity of Lanthanum, Neodymium and Samarium'. Further papers followed on the same sort of subject until 1935 when there is a change to an interest in neutrons, and the *Physics Review* in that year carried two of Libby's publications, 'Absorption and Scattering of Neutrons' and 'The Action of Neutrons on Heavy Water'. Throughout the rest of the 1930s Libby contributed a steady stream of work on both radioactivity and the characteristics and reactions of neutrons.

In 1941 the war claimed him, and he moved east to the Columbia University War Research Division. He produced one paper in 1942 and only one again in 1943; both concerned the hydrogen isotopes tritium and deuterium (heavy hydrogen). From 1943 to 1946 there were no published papers by Willard Libby. The reason is obvious.

In 1939 there could have been few, even among the élite of scientists involved in nuclear physics, better equipped than Libby to follow up Korff's discovery of high-altitude neutrons and the knowledge that neutrons in air formed radiocarbon. But at the end of the war no one had even yet found radiocarbon in nature, though Libby had by then formed the theory of radiocarbon dating in his own mind. He emerged from his wartime work as one of the leading nuclear chemists of the world – he had been Assistant Professor at the University of California just before the war started, he was made Associate Professor in the Department of Chemistry there when the war ended, but almost immediately he moved to the University of Chicago where he was appointed Professor of Chemistry in the Institute of Nuclear Studies and the Department of Chemistry in 1945. It was during his tenure at Chicago that he established the theory and practice of radiocarbon dating, though he returned to the west coast at U.C.L.A. in 1959, where he is now Professor of Chemistry and Director of the Institute of Geophysics and Planetary Physics.

As soon as he arrived in Chicago, Libby took up where he had left off at the start of the war; his first postwar paper, in 1946, was 'Atmospheric Helium Three and Radiocarbon from Cosmic

Radiation'. But it took him five years to show that his wartime theories were justified, five years to show that there was a radiocarbon clock, that it worked all over the world and that it could be used by archaeologists to give dates to the things they found by digging. It was not till 1950 that the 'clock that runs backwards' was in use for the regular telling of historical time, and it was not till 2 February 1951 that Libby could start producing the regular, published lists of the datings of archaeological materials which he still produces today.

Libby's theory was that, since radiocarbon was being made continuously in the high atmosphere and since no one had ever discovered it in nature, and since, presumably, the earth had been under bombardment by cosmic rays for many tens of thousands of years at least, then the radiocarbon must be disappearing by radioactive disintegration at exactly the same rate that it was being formed. In other words there must be an equilibrium between the rate of formation and the rate of disintegration of radiocarbon. Furthermore, this equilibrium must presumably have existed for thousands of years – that is to say there was exactly as much radiocarbon on earth at any time in the past as there is now. If radiocarbon entered living matter at all, there would have been exactly as much radiocarbon in a tree or plant or person living three thousand years ago as there is in a tree or plant or person living today.

There was good reason to believe that radiocarbon might enter living things. Carbon atoms do not float about by themselves singly in the normal course of things; they are invariably found joined to other elements in molecules, or to other carbon atoms in carbon molecules as in diamonds or graphite. Carbon is the essential constituent of living things – we, and all other life on earth, are carbon-based. It is also possible to find carbon in many non-living things, such as rocks and gases, but no living thing exists that is not dependent on and made up of very large molecules which always have carbon in them.

The radiocarbon made from nitrogen by the neutrons from the cosmic-ray collisions would almost certainly combine with the oxygen in the high atmosphere to form carbon dioxide, CO_2. The combination of carbon with oxygen to form carbon dioxide is a very common phenomenon – it is simply burning. But the carbon dioxide in our atmosphere is the basis of all food. Carbon dioxide is taken in by plants which also take in energy from the sun and use this energy to convert the carbon dioxide molecules into sugars and starches – the first of the big carbon-based molecules which are the feature of all life. Every other living creature lives off plants – the large animals and ourselves, obviously; but all the bacteria and micro-organisms and insects and earthworms live off the material of plants, whether alive or dead or rotting or after it has been turned into something else by a creature that has eaten the plant directly. In the seas and oceans the process is the same; marine plants, the tiny plankton, use photosynthesis to convert carbon dioxide into the big carbon-based molecules which are the stuff of life and all the fishes feed off the plankton or off each other.

The carbon dioxide of the atmosphere therefore enters all living matter through the plants. If the carbon dioxide of the atmosphere contains radiocarbon, it follows that radiocarbon must enter all living matter as part of the process of life.

But the moment that any creature dies, it stops taking in radiocarbon, or indeed any carbon at all, as a matter of general principle. For 'taking in radiocarbon' means only that among the many millions of atoms of carbon that a plant or animal absorbs and uses to build up its own structure, there are some atoms of the radioactive carbon isotope, carbon 14. A dead creature is of course in contact with carbon in the atmosphere, in the earth or in water – but the essence of death is the stopping of the metabolic processes, the chemical activities of a living body, and so the surrounding carbon – the ambient carbon – is not absorbed into a dead body's structures. (In certain places, where there is a great deal of calcium carbonate in the water or the soil, a dead body of plant or animal may be 'contaminated' by ambient carbon as it lies dead, but this is a special circumstance which can be allowed for by an archaeologist wishing to use the carbon-dating technique.)

In general, however, a dead creature stops taking in radioactive carbon when it dies. The radiocarbon in the creature or plant at the moment it dies will steadily disintegrate, emitting radioactivity as it does so, and turn into atoms of nitrogen. Therefore the longer a plant or animal has been dead, the lower the proportion of radioactive carbon it will contain. If we know how much radiocarbon it contained when it died, and we know the rate at which radiocarbon disappears to become nitrogen, and we measure how much radiocarbon the dead material contains at the present moment we can calculate how long it is since the material became dead.

The moment of death of a plant or animal therefore sets a clock at zero. By measuring the amount of radiocarbon it contains at any time after death we can read off directly the amount of time since it died. This applies to any part of a plant or animal – it applies to the skin of an animal which has later been turned into leather; it applies to the leaf or stem of a reed which has been turned into paper; it applies to the trunk of a tree which has been used in a building; and it applies to the bones of a buried human body.

In the use of phrases such as 'the amount of radiocarbon' and 'how much radiocarbon' in the preceding paragraphs there has been a slight oversimplification. An elephant contains more radiocarbon than a mouse in total. But Libby's assumption was that a living elephant contains the same proportion of radiocarbon atoms to ordinary carbon atoms as a mouse because it appears that throughout the world there is an equilibrium between radiocarbon atoms and ordinary carbon atoms – the number of radiocarbon atoms being created by cosmic rays crashing into the nitrogen of the atmosphere equalling the number of radiocarbon atoms returning to nitrogen by their own disintegration. The evidence for this bold assumption was that it was known that radiocarbon atoms were being created by cosmic rays regularly, yet there was no trace of a build-up of radiocarbon atoms in living things.

Another assumption implicit in the theory was that radiocarbon was evenly distributed throughout the earth – which, in practice, means that a mouse in New York State contains the same proportion of radiocarbon atoms to normal carbon atoms as a mouse in Egypt or New Zealand. It was by no means inconceivable that more cosmic rays might strike the atmosphere near the poles of the earth than near the equator, and thus one could visualize a situation in which, though there was an equilibrium between newly formed radiocarbon and disintegrating radiocarbon at all places, the level of that equilibrium might be different from place to place. This assumption could only be checked by direct measurements, but if it was not valid the radiocarbon clock, of which Libby dreamed, would not be universally applicable.

Libby's theory also depended on the assumption that the radiocarbon equilibrium had existed, and existed at its present level, throughout the tens of thousands of years that the radiocarbon clock would, he hoped, measure.

There were, in fact, four major assumptions in Libby's theory. First, he assumed that the ratio of radiocarbon to ordinary carbon in the atmosphere was the same at all historic times; secondly he assumed that a living creature will, at any time, contain within itself that same proportion of radiocarbon to ordinary carbon atoms that the atmosphere contains; thirdly that, once a creature has died, the ratio of radiocarbon to ordinary carbon atoms will change only by the disintegration of radiocarbon atoms; and fourthly that the rate of decay of radiocarbon atoms by disintegration, can always be accurately measured or calculated.

These were four very large assumptions, especially at a time when no one had ever even found any radiocarbon in any living thing, and this was obviously where Libby had to start, in order to turn his theory into a useful tool for archaeology.

The hunt for natural radiocarbon began in 1946. Since no one had ever discovered any natural radiocarbon in living creatures up till then it was reasonable to assume either that it did not exist, or that the instruments of the time were not sufficiently sensitive to pick up the low levels of radioactivity that would be involved. Manmade radiocarbon was known, however, at the time, because it had been produced artificially in the nuclear accelerators and nuclear laboratories where the science behind the atom bombs had been carried out. Libby knew therefore how much radioactivity and what energies of disintegration he was looking for. (When a radioactive atom 'disintegrates' and changes into another type of atom, emitting radioactivity, the radioactivity has a certain level of energy which is specific to the precise reaction involved, so that, by measuring the energy of the disintegration accurately, the scientist can tell exactly which sort of atom has disintegrated.)

The only method by which Libby could find if there was any such thing as natural radiocarbon in living matter was to take some carbon from living creatures and try to concentrate the atoms of carbon 14 which he guessed were there. The process of isotope separation or isotope enrichment – the obtaining of a material in which atoms of a particular isotope were more common than normal – had been intensively studied during the wartime nuclear weapons programmes. Indeed the very existence of some nuclear weapons depended on success in concentrating or 'enriching' the rare, but fissionable, uranium 235 isotope. But no one had ever bothered to concentrate carbon 14, and Libby admits that he and his colleague, E. C. Anderson, were 'stumped for a time'. Then he remembered a wartime colleague, Dr A. V. Grosse. He had been a scientist at Temple University but he had gone to work with the Houdry Process Corporation at Marcus Hook, near Philadelphia. He was at this time working on a process to concentrate carbon 13 atoms from ordinary carbon, with the object of getting carbon 13 in sufficient quantities for use in cancer research where the unusual carbon atom could be traced in its progress through living bodies because of its difference from ordinary carbon atoms. Dr Grosse had constructed a large and expensive plant for the concentration of carbon 13 and this plant used as its raw material the common gas, methane, which is the combination of carbon and hydrogen atoms with the formula CH_4, i.e. a combination of one carbon with four hydrogen atoms in each molecule.

Dr Grosse agreed to try to enrich carbon 14 in his isotope-separator – if indeed such atoms existed in ordinary living matter. It was up to Libby to find the source material and this had to be methane to suit the machine. Most methane comes as 'natural gas' from the oil and gas fields – whether this has been formed from living matter or not, it has been stored for so many millions of years underground and out of contact with the air that any radiocarbon it may originally have contained has long since totally disintegrated and disappeared. Libby found his source of methane in the Baltimore sewage works – in plain words he got it from the excretions of the citizens of Baltimore. He knew that, if there is natural radiocarbon in all living things, it will, of course, be in all the meat and vegetable foods of human beings and, correspondingly, in the carbon contents of their excretions. The methane gas produced in the sewage works would be very much 'living' carbon in his sense of the word. Libby himself described the results in these words:

Well, Dr Grosse, after obtaining the sewage methane, proceeded to enrich it to various degrees (as measured by the carbon-13 enrichment) and Dr Anderson and I excitedly put the enriched methane in our proportional counter and took the counts behind the heaviest shield we could assemble, since the counter was responding in the main to laboratory radiation present in the building and equipment, and to the cosmic radiation reaching the Earth's surface at Chicago. Strangely enough the whole thing worked and we did find about the anticipated carbon-14 concentration as a small additional count-rate for the enriched methane as compared with the rate for unenriched methane or for petroleum natural-gas methane. Further confirmation came when the carbon dioxide formed by burning the methane was found to be radioactive as well and to display a radiation with an average penetrating distance equivalent to 2.5 cm. of air just as is true of man-made radiocarbon.[1]

So the bowels of the citizens of Baltimore unwittingly provided the first sample of natural radiocarbon to be discovered. The first and most important of Libby's theories proved to be correct – natural radiocarbon did exist in living organisms.

These words of Libby's were originally spoken to a gathering of scientists at the Royal Society in London during a 1970 symposium on radiocarbon dating. They assume, therefore, an understanding of several things. There is, for a start, the method of counting radioactive disintegrations. It is widely known that a Geiger counter is the machine for measuring radioactivity, and Geiger counters were used in much of the later work of measuring archaeological samples. But a Geiger counter, or Libby's 'proportional counter', or the 'scintillation counters' that are used nowadays, all work on the same principle. The radioactivity released by a disintegrating atom appears in the form of one of the basic nuclear particles – an electron in the case of radiocarbon – with a specific energy. A particle of this sort travelling through a gas, 'ionizes' the atoms along its path, which means that it changes the electrical properties of these atoms. The more energy the particle has, the more atoms it will affect and the greater will be the electrical disturbance it causes. All forms of radioactivity counters detect these electrical changes caused by the flight of the particle and measure the size of the disturbance, and the essence of a counter is that it magnifies the very small disturbance created by the particle to such a size that it can be 'spotted' and measured; this is normally done by some means such as putting the gas inside the counter under a high voltage, so that the disturbance caused by the particle appears as a large pulse.

One of the minor 'bonuses' of using carbon for dating is that carbon is a constituent of many common gases such as methane or acetylene, or carbon dioxide, the normal product of burning any carbon-containing object, and therefore the carbon sample whose radioactivity is to be measured can be converted into one of these gases and used as the gas inside the counter, measuring its own radioactivity. This gets rid of many problems of contamination which would disturb results if a sample had to be introduced into the gas of the counter.

Libby also mentions the 'count-rate' which turned out to be roughly what he 'anticipated'. In fact his theory had predicted the amount of radiocarbon he should find in living things. It was known from the early balloon-experiment days before the war that the rate of bombardment of the atmosphere by neutrons coming from cosmic-ray collisions worked out at an average of two neutrons every second on each square centimetre of an imaginary shell surrounding the earth. That meant that two atoms of radiocarbon were being formed each second in the column of air above each square centimetre of the earth. Now if one imagines all the carbon atoms on the earth separated from the other atoms and dumped into little piles on each square centimetre of the earth, it can be calculated that there will be about 8.5 grammes of carbon in each pile. So each pile of eight and a half grammes of carbon is having two extra radiocarbon atoms added to it each second. But, since there is no increase of radiocarbon on the earth, two of the radiocarbon atoms in each pile of eight and a half grammes of carbon must be disintegrating each second. So if the scientists puts eight and a half grammes of 'living' carbon into his counter he ought to get two counts per second, which is 120 a minute. In fact Libby's counter could take only about 5 grammes of carbon in the form of methane, so he expected about 75 counts a minute without enrichment of the sample. (The 8.5 grammes of carbon on each square centimetre of the earth is almost all contained in the oceans. Only one third of a gramme is provided by living creatures including all the plants, and just over one tenth of a gramme is in the carbon dioxide of the atmosphere.)

Libby also talked easily to the Royal Society about 'shields'. The problem of shielding, or screening, the counters was, in fact, his first and biggest problem once he had established that natural radiocarbon did exist. Radiocarbon dating would never be a practicable tool for archaeologists if isotope separators costing tens of thousands of dollars had to be used all the time and if there was no way of getting the counting methods down to a reasonable operating cost. Yet there is a considerable amount of 'natural radioactivity' everywhere on the earth. Most of this comes from the radioactivity of uranium, thorium and potassium, ordinary chemical elements found everywhere in small quantities. Most of the remainder of the radioactivity comes from cosmic rays bombarding the earth; the same cosmic rays that produce the neutrons that make radiocarbon in the atmosphere also produce many other particles in their various collisions with the gas atoms of the atmosphere and these other particles, technically mesons, shower down on us all the time quite harmlessly. But the total of all this 'natural radioactivity' gave Libby 500 counts a minute in his Chicago laboratory counter. He could never measure small variations on the mere 75 counts a minute provided by living radiocarbon amid this deluge of other radioactive particles. So he shielded his counter with a layer of iron eight inches thick all round. This screened out all the particles from the uranium and thorium and potassium and reduced the count of unwanted particles to about 100 a minute. But it could not stop the cosmic-ray mesons – no practicable amount of any material on earth can stop mesons.

The answer to the meson problem had, therefore, to be found in ingenuity rather than in strength. What Libby and Anderson did was to surround their counting chamber, inside the thick iron screen, with a dozen Geiger counters packed close so that they touched each other. The mass of Geiger counters was connected in such a way that as soon as any one of them recorded the passage of a particle – one of the mesons that had penetrated the iron screen – the counting apparatus in the main chamber was switched off. This switch-off lasted only one thousandth of a second but it was enough to eliminate the mesons from the count and to reduce the unwanted counting in the main chamber, the so-called 'background' count, to between one and six pulses a minute. Though there have been many improvements in the twenty-five years that have followed Libby's early experiments, in principle the modern counters in museums and departments of archaeology all over the world still follow Libby's main ideas.

Libby had by this time solved the problem of measuring the very low level of 'natural' radiocarbon radioactivity in a practical and not over-expensive way. So now he could set out to test the second assumption of this theory, which was that throughout the tens of thousands of years of man's history the winds, the general movements of the earth's atmosphere and the currents of the oceans had spread the radiocarbon that was being formed in the outer layer of the atmosphere evenly throughout the world.

The idea was simple – they would collect living material from all over the world and measure its radiocarbon content, and they hoped that all the measurements would be the same. Much of the detailed work here was done by the younger man, E. C. Anderson, and the results formed his doctoral thesis at the University of Chicago. At this stage, too, it was necessary to get help and funds from outside bodies – the Guggenheim and Wenner-Gren Foundations, the American Geological Society and the U.S. Air Force all came up with help. They decided to concentrate on wood samples and they enlisted the help of Donald Collier of the Chicago Natural History Museum. With his help they got samples of ironwood from the Marshall Islands exactly on the equator, and pinewood from Sweden, oak from Palestine and eucalyptus from New South Wales in Australia as well as samples from Japan and West Africa, and the mountains of Switzerland. Libby picked up local elmwood in Chicago and friends sent honeysuckle leaves from Oak Ridge in Tennessee. They got samples of spruce from the far north of Alaska and beechwood from Tierra del Fuego at the opposite, southern, tip of the Americas. They got pine twigs and needles from 10,000 feet up Mount Wheeler in New Mexico, and Donald Collier provided wood from the high Andes in Bolivia. North African briar and lumps of unidentified wood from Panama and Persia completed their collection. They could not get wood from the plantless Antarctic, so they got seal oil that had been collected by Admiral Byrd's South Pole Expedition. All these samples had one feature in common: they were first cut at about the beginning of the present century. This made samples rather harder to come by but it was essential to avoid any suggestion of contamination by nuclear explosions or even by the enormous output of carbon dioxide which our twentieth-century civilization pours into the atmosphere.

The results of measuring the natural radiocarbon in all these samples were immensely encouraging. They all gave exactly the same reading within the limits of accuracy of the counter. Libby's next assumption had been proved. They had also, in the process, added several major pieces of knowledge to the general corpus of scientific fact. They had shown that radiocarbon does exist in living matter throughout that world and they had shown a very important fact about the circulation of the earth's atmosphere, namely that if you put something into the atmosphere it will eventually be spread out evenly all over the globe. Several other experimenters in different laboratories confirmed Anderson's results in the following years.

From this point Libby could go on to actual dating. He had shown that radiocarbon exists at the same level throughout living matter, in other words his clock idea was based on reality. Now the questions arose of how to tell the time, and whether the clock actually ticked on through the centuries.

Wartime nuclear work had taught the scientists a great deal about the disintegration of unstable radioactive isotopes and atoms. It had been shown that there was no means of telling when any one particular radioactive atom would disintegrate. But it had equally been shown that, although the disintegration of any one atom was a purely random event, if you took enough atoms of the same sort, the combination of purely random disintegrations produced a regular flow of disintegrations. Some radioactive materials disintegrated very quickly while others disintegrated only very slowly. The rate at which different radioactive atoms disintegrate is described as their 'half-life'. This means that, although you can never say when one particular atom will disintegrate, if you take a block of pure radioactive substance then half of that block will have disintegrated and turned into something else in the specified time. Some radioactive atoms have half-lives of millions of years, others have half-lives of only fractions of a second.

The half-life of radiocarbon had been calculated from manmade samples at 5,568 years, and this was the figure Libby used at that time, although he and others have, in later years, made more accurate calculations and measurements. This means that if you took one gramme of pure radiocarbon, then in approximately 5,600 years you would have only half a gramme left, and the rest would have disappeared as nitrogen 14. But it does *not* follow that in 11,200 years you would have none left at all. In the second

5,600 years exactly half of the half gramme would disintegrate and you would be left with one quarter of a gramme after 11,200 years. 5,600 years later still there would be one eighth of a gramme remaining and so on. In purely mathematical terms the process never ends, but in practice the disintegrations can eventually no longer be counted by our instruments.

In the practical terms of Libby's laboratory counting, what this meant was that, since he had got a count-rate of 75 disintegrations a minute from the radiocarbon of a five-gramme sample of modern living material, a five-gramme sample of ancient material that gave him 37½ counts a minute would be calculated to be 5,600 years old. At the time of the sample's life it would have had enough radiocarbon in it to give 75 counts just as a modern sample would. If the sample is giving only half that count nowadays it must have lost half its natural radiocarbon and must therefore be 5,600 years old. Correspondingly a sample giving 18¾ counts a minute should be 11,200 years old and a sample giving only three-quarters of a count per minute should be 56,000 years old. Of course the counting measurements were not performed for just a single minute, but were carried on for several hours, perhaps twenty-four hours, and the counts-per-minute calculated afterwards.

With this theory in mind, Libby set about finding whether the clock really ticked on in this way. The obvious thing to do was to get a variety of ancient specimens, measure their radioactivity, calculate their dates from this, and then see whether the results corresponded to what the archaeologists and historians said the dates should be. But, whereas collecting specimens of fifty-year-old wood from around the world is not too difficult, museum curators and collectors are not going to be willing to give up valuable specimens to be burnt and destroyed by a 'mad nuclear scientist' just to prove some crazy theory that he can give you a date for anything you dig up regardless of what it is or where it's found. At this stage Libby had a stroke of luck. He took onto his team a young physical chemist from Princeton, Dr J. Arnold. (Both Libby and his colleague, Anderson, were physical chemists, too.) It turned out that Dr Arnold's father was a lawyer and a keen amateur archaeologist. It was agreed that the only way for the physical chemists to get the co-operation of archaeologists was to approach them officially and ask for their help and advice. So Libby went to the American Archeological Association and the Geological Society of America, told them what he was trying to do, and asked them to give him a committee of experts to advise. There was no hesitation in appointing three archaeologists (Frederick Johnson, Peabody Museum, chairman, Froelich Rainey, Philadelphia Museum, and their old friend Donald Collier of Chicago) and one distinguished geologist (Richard Foster Flint of Yale). In Libby's words: 'These gentlemen did it right.' But first they showed the scientists that things were not going to be as simple as they had expected:

> The first shock Dr Arnold and I had was when our advisors informed us that history extended back only to 5,000 years. We had thought initially that we would be able to get samples all along the curve back to 30,000 years. [Author: the curve being the line on graph-paper that Libby had drawn to show what the radioactivity readings should be for samples of different ages], put the points in, and then our work would be finished. You read statements in books that such and such a society or archaeological site is 20,000 years old. We learned rather abruptly that these numbers, these ancient years, are not known accurately; in fact it is at about the time of the first dynasty in Egypt that the first historical date of any real certainty has been established. So we had, in the initial stages the opportunity to check against knowns, principally Egyptian artefacts, and in the second stage we had to go into the great wilderness of prehistory to see whether there were elements of internal consistency which would lead one to believe that the method was sound or not.[2]

The words are Libby's, spoken when he received his Nobel Prize in 1961. At that time, more than ten years after the work was started, some of the problems set by his advisory committee, problems 'in the great wilderness of prehistory', were still being studied. It was very far from the simple collection of samples and comparison with a curve for which the physical chemist had hoped.

Still the essence of the first task facing them was the same – to test the radiocarbon dating technique against antique samples of known age – even though the stretch of 'historic' or 'known age' times was so much less than Libby had expected. They turned to the Oriental Institute of their own university, the University of Chicago, which kept the rich Breasted Collection of Egyptian antiquities. From Professor John Wilson they got permission to take and burn small samples of his priceless treasures, and he helped them to get yet further samples from his colleagues.

They got wood that had been found in the underground tombs of King Zet and the Vizier, Hemaka, near Saqqara, one of the oldest known Egyptian sites. These tombs are of the First Dynasty and therefore can be dated pretty accurately by historical methods to around 3000 B.C., pretty nearly five thousand years before our own time. Then there was a sample of cedar wood from the upper chamber of the Pyramid of Sneferu. Perhaps the sample they prized most was the small portion of the deck timber of the Funeral Boat of Sesostris, which they were allowed to take. This boat is one of the treasures of the Chicago Museum of Natural History; it is twenty feet long and six feet wide, complete with paddles, and it was placed in the tomb of Sesostris III, the Pharaoh, presumably to carry his soul across the waters to the other world. The historians and Egyptologists date it to 1800 B.C. Wood from an Egyptian sarcophagus of about the same date completed the Egyptian sample of 'knowns'.

More recent 'knowns' were provided by wood from the floor of a central room in a palace at Tayinat in north-west Persia, which was known to have been burnt down in 675 B.C.; the linen wrapping of one of the Dead Sea Scrolls found in Palestine – a version of the Book of Isaiah which can be dated to within a hundred years of the start of the present era; and an overcooked, carbonized, baker's roll of bread buried in the volcanic ash that overwhelmed the city of Pompeii in A.D. 79 precisely.

An entirely different type of sample was brought in to give a cross-check on these relics of the old civilizations around the Mediterranean. Samples were also taken from the cores of several of the giant redwood trees of North America. One of these was the famous 'Centennia Stump' which was felled in 1874 and which, by counting the tree-rings, has been shown to be about 2,930 years old. It fitted, in time terms, into the gap between the latest of the Egyptian specimens and the earliest of the Middle East specimens. Other samples of the heart-wood of the giant sequoia trees were provided to cover the period of around A.D. 1000.

The immediate results of testing these samples of known age by measuring their radiocarbon content were triumphant. Every single sample was given an age closely corresponding to the age the historians allotted to it. Of course every point did not land exactly on the theoretical graph that Libby had drawn. Some points were slightly to one side, others were off in the opposite direction. Radiocarbon seemed, in particular, to make Egyptian objects slightly 'younger' than they really were, but only by a few hundred years in the most extreme cases, and for many other samples it gave a perfectly correct age. But any variations were well within 'experimental error' and there were many facets of the technique which could clearly be improved – the calculated half-life of radiocarbon, for instance, might have to be measured more accurately. On the other hand Libby had taken the precaution of obtaining extra quantities of some of his samples and submitting them to be measured for radiocarbon content by Miss Elizabeth Ralph at the Philadelphia Museum. Her results were closely in accord with his own.

So it was clear that radiocarbon dating could measure the time at which a creature, plant or animal had been alive, and do it accurately back through history, at least as far as could be checked by objects of known date. This work had been concluded and published in 1949, and it is from 1949 that we must count the universal clock as being available.

But the committee of advisers insisted that the clock must be tested on a worldwide scale and in the 'great wilderness of prehistory'. They thought up a dozen different projects in which the radiocarbon method could be used for dating prehistoric sites and objects with a view to seeing whether the new technique was reliable, whether it was internally self-consistent, and whether its results were reasonable dates in the contexts of prehistory – there might be thousands of years of various cultures or civilizations in many different parts of the world where the archaeologist and prehistorian could not give dates but where they knew quite definitely that one stage of a culture came before another, and where, therefore, the reasonableness of radiocarbon dating could be tested.

Of all the tests set for radiocarbon dating the most understandable, and in many ways the most crucial, was the one relating to the final Ice Age. It was argued that the great ice sheets which covered much of North America and northern Europe must have swept away any traces of human occupation of the countryside when they came. Libby wrote of this experiment,

> We were afraid that we should find man older than the last Ice Age and we had agreed that this would constitute sufficiently conclusive evidence to discredit the whole method; we felt that glaciers sweep very clean and there should be no evidence of earlier human occupation left. So in England which was completely glaciated there should be no evidence of human beings older than the time of the last glacier.[3]

Obviously, this statement needs qualification in the broadest terms of human history, but Libby's meaning is clear.

Once again the clock passed the test highly successfully and even revealed a good deal of new information in doing so. It showed that the last ice sheet came down over North America and northern Europe at almost exactly the same time, about 11,400 years ago. Further, it showed that all the sites of human occupation in England, Ireland, Scandinavia and north Germany are dated to 10,400 years ago or later. Man's occupation of all these areas is plainly subsequent to the last ice sheet. Yet at the same time the radiocarbon dating showed clearly that south of the ice sheet, round the Mediterranean, man had lived continuously both before and after the last Ice Age – the famous cave paintings at Lascaux in France, which was not reached by the glaciers, date to five thousand years before the last Ice Age.

In North America the situation is rather different. The last Ice Age occurred at the same time as in Europe, radiocarbon dating showed this clearly. But at that time there was considerable dispute among the archaeologsits themselves about the date that could be given to 'the first American'. Since then it has been clearly established that Central and South American sites show man occupying these areas at 20,000 years ago and more. But at the time when the radiocarbon clock was being put through its tests there was much argument between the archaeologists as to whether certain sites were 'human' or not. Libby's radiocarbon results showed that a very large number of North American sites of archaeological interest dated to 10,400 years ago and very few indeed dated older than that. There is a clear discontinuity on a table which compares the number of 'human' sites with the age according to radiocarbon and this discontinuity occurs at precisely the time when the Ice Age glaciers swept down. This was quite satisfactory enough to establish the usefulness of the radiocarbon clock, but its exact meaning archaeologically is a matter on which Libby does not commit himself. He merely remarks,

> Most of the sites that are older than 10,400 years are equivocal in one way or another, at least to the chemist or physicist who overhears the archaeologists arguing about them. We have noticed that there is a considerable unanimity of opinion about American sites of 10,400 years or younger being human sites; whereas there is a considerable tendency for discussion and debate for the older sites.[4]

So by 1950 the 'clock that runs backwards' was established scientifically. It cannot tell whether sites are 'human' or not, but it can tell what dates the sites were occupied back as far as 50,000 years ago and there it reaches its limit for most practical purposes.

The radiocarbon clock does more than just tell the archaeologist the date of something he has found. It enables him to relate an independent site or an indigenous culture to what was happening in other parts of the world at the same time – it is a universal clock. More than that, however, it has changed the whole practice and profession of archaeology by forcing every archaeologist, wherever he may be digging, into an awareness of the importance of the laboratory sciences. And twenty years after it was proved to be a practical tool the results provided by the clock forged a revolution in our thinking about much of prehistory.

1. W. F. Libby, 'Radiocarbon Dating'.
2. W. F. Libby, *Nobel Prize Address.*
3. W. F. Libby, 'Radiocarbon Dating'.
4. ibid.

5

The Clock Reset

The first great triumph of radiocarbon dating was to fix the time-scale and dating of the Neolithic Revolution in the Middle East. It took about twenty years of excavation and dating to show that the whole process which goes under the package name of Neolithic Revolution had taken longer, and happened earlier than anyone had previously believed. It emerged in the course of this work that the revolution had not been one single event but a complex of many different developments such as the domestication of several different sorts of plants and animals, the discovery of pottery, the start of settled life in villages and towns, the creation of new religions, the increase of trading and the increase of populations. These differing developments occurred in many different places and probably independently. In some places they occurred in a different order from that discovered in other places. Many scientific techniques other than radiocarbon dating have been used in elucidating the picture (and these will be dealt with in later chapters). The process is by no means complete even now, but the main outlines of the dates of the Neolithic Revolution in the Middle East had been established by 1969.

The first date ever obtained for a Neolithic site anywhere in the world came in 1948, when Professor R. J. Braidwood, of the University of Chicago, found material which was dated to 4800 B.C. in the Neolithic village of Jarmo in the foothills of north-eastern Iraq. The importance of this achievement was simply stated by Kathleen Kenyon, the excavator of Jericho, 'Until then all dates has been guesswork.' She said this in 1969 when she gave the Presidential Address to the Anthropology Section of the British

Association. The rest of her address was devoted to showing that there had been a 'a revolution of the Neolithic Revolution' in the twenty years that followed the production of that date. By 1969 radiocarbon dating had established that the Neolithic Revolution in the Near and Middle East occurred broadly between 10,000 B.C. and 4000 B.C.

Similarly, radiocarbon dating has also shown that in Mexico and Central America a development comparable to the Neolithic Revolution, achieved independently by the men who lived in those parts of the world, took about the same length of time – from rather before 5000 B.C. to the early centuries A.D. So radiocarbon dating gave the first-ever reliable measurement of the times when these supremely important developments in human history took place and, perhaps even more important, gave the first reliable calculations as to how long it took human beings to move from what is essentially 'savagery' to what we call 'civilization' – and the move from savagery to civilization is another definition of the Neolithic Revolution.

But if radiocarbon dating had been triumphant in fixing the timing of the broad sweep of Neolithic Revolutions it seemed less successful when more detailed questions were asked. There was, for instance, the problem of the time and length of the 'Neolithic' period in central and northern Europe; there were two schools of thought among archaeologists even before the radiocarbon method was invented, the 'short-chronology' school and the 'long-chronology' theory. The short chronology placed the Neolithic in Europe from just after 3000 B.C. to around 1700 B.C. It was based on several specific archaeological 'finds', which were compared with corresponding objects found in Egypt and Mesopotamia where they could be pretty accurately dated from historical records. The situation that then arose after the discovery of radiocarbon was described by E. Neustupny, the Czech archaeologist.

This chronology (the short chronology) seemed to be well founded and most leading European archaeologists accepted it and used it in their publications.

When the first radiocarbon dates appeared in the 'fifties' they showed figures higher by almost 1,500 years. This seemed unacceptable to many archaeologists, who looked upon the radiocarbon method with distrust, and some even began to contest the very foundations of the method.

The opponents of the radiocarbon dating method in central and south-east Europe (mainly Milojcic) succeeded in evoking an atmosphere of distrust, in which most archaeologists did not accept the existing radiocarbon dates and did not collect samples on which new dates could be obtained.

The 'short' historical chronology survived successfully, along with the increasing number of radiocarbon dates obtained mostly in Groningen and Berlin. There were however a few archaeologists who argued in favour of the high (or 'long') chronology on archaeological grounds, and many others who accepted radiocarbon dates.[1]

The resolution of this great controversy by the overwhelming evidence of radiocarbon in favour of the long chronology has been one of the most important developments of archaeology and prehistory of the last few years. Indeed, it can be said to have been largely responsible for sparking off that revolution in archaeology, the 'new archaeology', which has been the most significant development of the 1970s in this area. But the very fact that there could be controversy over radiocarbon dates points to the struggle and confusion of the early days of radiocarbon dating.

One of the biggest sources of trouble came in comparing radiocarbon dates with the dates provided by the Egyptologists from the huge mass of well-preserved Egyptian material from the years of the pharaohs. Here the new scientific methods were plainly open to some very destructive criticism. Broadly speaking, the radiocarbon method gave dates that were pretty regularly two hundred or three hundred years too low. For instance, the radiocarbon date of the wood from Tutankhamun's tomb was about 1050 B.C., whereas the historical date is roughly 1350 B.C. The further back in Egyptian history, the larger became the error, so that it was often as much as five hundred years too young by radiocarbon reckoning for objects from 'The Old Kingdom' which were known historically to date from just after 3000 B.C.

The work of checking radiocarbon dating against the well-known Egyptian chronology was never performed systematically in the first twenty years of the history of radiocarbon, and this was for a variety of reasons. First, the Egyptologists had no good reason to be interested in radiocarbon datings; the chronology of Egypt of the pharaohs is not, in fact, known as accurately as many Egyptologists seem to imply, but there are three or four crucial dates in Egyptian written (or carved) records giving the heliacal rising of the Sothis star, which we know as Sirius. Calculations backwards from modern astronomical data enable us to fix these times within a few years. Whatever may be the details of the rival theories of professional historians of ancient Egypt, there are few important dates in the time of the pharaohs that are not known within about twenty years either way. Radiocarbon dating cannot match this accuracy for a specimen three thousand years old – the best laboratories can only give a radiocarbon date within about a hundred years at that distance away. So there was no compulsion on the Egyptologists to date their materials by radiocarbon methods when their traditional documentary and stylistic evidence would give them a more accurate result. From the scientists' point of view material was more difficult to come by since organic matter, wood, leather, papyrus, necessary for radiocarbon work was normally the precious possession of the museum that held it, and to cut away bits of these priceless relics and to destroy them in the process of radiocarbon work was unthinkable. By 1969 fewer than a hundred samples of ancient Egyptian material had been dated by radiocarbon methods. Yet it was quite clear that, even allowing for the minor uncertainties in Egyptian historical dating and the greater uncertainties in the laboratory results for the fifty samples that had been dated by radioactivity, there was this serious discrepancy of between two hundred and five hundred years in the radiocarbon dates of Egyptian samples.

It is one of the problems of hindsight and history that, when we have identified the crucial moment, the exact point of time at which the tide turned, we ignore the later struggles of the forces or ideas that were doomed to eventual defeat. In military terms, mopping-up operations are distasteful to conduct, expensive in manpower, and unworthy of record. Yet the fact is that it took twenty years before radiocarbon dating became accepted as an established technique which, if treated properly, could be regularly relied upon by the archaeologist to solve the dating problems of a site.

There are probably two main reasons for this. The first is that radiocarbon dating crossed the traditional boundaries between academic disciplines – indeed it crossed the great divide between arts and sciences. The second reason is that it took the scientists, and those who favour a scientific approach to archaeology, twenty years to establish an agreed set of satisfactory and well-tested techniques for radiocarbon dating and to find an independent checking system by which the radiocarbon findings could be calibrated.

There is always opposition, or at least guarded neutrality, when a new technique is applied outside the borders of the discipline that invented it, and all the more so when there are social consequences of this transgression across borders. Techniques such as the ultracentrifuge or the automatic scintillation counter attract no criticism since they are confined to the sciences which first brought them into use. No one criticizes even the computer when its operations are confined to calculating the speeds and energies of sub-nuclear particles, but it becomes a very controversial machine when it is used to study the authorship of New Testament texts, for then it impinges upon the religious beliefs of many people and upon the professional lives of the men who devote their time to biblical studies.

Some archaeologists refused to accept radiocarbon dates. The attitude of the majority, probably, in the early days of the new technique was summed up by Professor Jo Brew, Director of the Peabody Museum at Harvard, 'If a C14 date supports our theories, we put it in the main text. If it does not entirely contradict them, we put it in a footnote. And if it is completely "out of date" we just drop it.'

The reluctance of many archaeologists to accept radiocarbon

dates if the dates showed signs of overruling current theories was much increased by the fact that there was often disagreement between the results obtained by the different laboratories that took up the new technique. Right at the start it was clear that different laboratories were producing different dates for the same specimen, and at the worst a single laboratory might get different dates from two counts on the same specimen. In any laboratory experiment it is usual to get slightly different readings from the instruments even though conditions appear to be identical. The scientist is perfectly accustomed to getting out of this sort of difficulty by taking a large number of readings on the same specimen, or with the same instruments, and then calculating the mean of all the readings to give a final result. However, the full width of variation of the results must also be published to show the precise amount of accuracy that has been achieved.

Taking several readings from a single specimen would have been normal practice in any case for dealing with the phenomenon of radioactivity, which is known to be a random, and not a regular, emission. Nevertheless it took a good many years for the chief radiocarbon-dating laboratories to get their readings into line with each other. This was largely a matter of setting up mutually acceptable standards of calibrating the instruments, working out satisfactory cleaning methods and making allowances for the background radiations, which vary from one part of the world to another. Most laboratories had special problems of their own to overcome – and the example of the Cologne Laboratory in Germany is worth noting. As late as 1968 they were getting results which regularly differed by 100 up to 250 years from those provided by other near-by universities such as Hanover, Berlin and Heidelberg. To all appearances their laboratory set-up was perfectly standard and satisfactory, and it took many weeks of work and apparently fruitless checking back of results before they found that there was a very slight seepage of gas into the laboratory building. This gas came from the city sewers and it contained radioactive components and was therefore distorting the results of the laboratory's work.

Undoubtedly the biggest source of potential error in radiocarbon dating, when once the laboratory has been properly calibrated, comes from contamination of the specimen. Throughout the process of excavation and examination by archaeologists on the site, throughout the journey to the laboratory, the specimen or sample is threatened by contamination from present-day, living carbon-containing substances or even from the carbon gases of the atmosphere. During the centuries that the sample has lain underground it can have become contaminated at any time by water containing carbonaceous chemicals or by plant roots growing through or round it. Washing techniques have therefore been an essential part of the carbon-dating laboratory's work right from the start. Most of the dating-specialists are now convinced that washing techniques are successful and that contamination can be countered in this way. Even when washing or cleaning fails to eliminate all contamination its presence can be spotted by splitting the sample into several separate moieties, giving different cleaning treatments to the various moieties, and comparing the measured dates. This then allows proper compensation to be given for the contamination.

Libby himself has strong and interesting views on the matter of cleaning. He himself posed the question, 'Why are ancient materials cleanable?' and answered it:

> Perhaps the most surprising aspect of our experience with radiocarbon-dating is that most ancient materials can be cleaned, so that only the original carbon remains. Before considering possible explanations and the limits that may exist to the cleaning process, let us think about the seriousness of contamination. At 57,300 years the radiocarbon left is expected to be 1:1,024 of that originally present. So modern carbon present to only 0.1 percent in much older material would give a false age of this magnitude. Thus we see that the question of contamination is most important for just those materials which have been exposed for the longer periods and that the true limit of the radiocarbon dating method may lie as much in the cleaning of the material as in the measurement of the radiocarbon radiation.

What then is it that keeps ancient organic matter separate from contamination, at least on the molecular scale, so that chemical and physical cleaning suffice? Well, the answer is quite obvious and consists of the enormous differences between the chemistries of life and death. The chemistry of life is first the chemistry of photosynthesis, the conversion of carbon dioxide, water salts and sunlight into starches, sugars, cellulose and many other substances, and then the chemistry of animals and insects living off the photosynthetic products, water and salts, producing the amino-acids and proteins, nucleics acids and bone, shell, and teeth.

In death all the magic has departed and everything runs downhill to carbon dioxide and water and sea-salts again, to be ready for the next cycle. None of the key molecules made in life are made in decay, so it follows, at least in principle, that one can always distinguish between the original matter produced during life and that produced during decay. In the end one may be forced to chemically separate the protein or cellulose from the simpler molecules produced during decay, since, indeed, some of the decay-products, such as amino acids from protein hydrolysis may be identical with some of the intermediates in the synthesis of the gigantic molecular structures of living beings. The point is, however, that this is usually possible and frequently not too difficult to do.

So with due precautions against various obvious contaminants – including the dangerous approach of the museum curator with his pot of molten wax or bottle of shellac – the scientists of the radiocarbon laboratories have managed to date bone, using its protein only, and teeth. Wood is one of the commonest materials to be dated because all tests so far show that although the living sap soaks even into the dead heartwood the two chemistries, of death and life, keep the constituents apart and the dead is not contaminated by the living. Shells and pearls can likewise be dated, since the material is calcium carbonate formed by the living processes of the creatures – though here the danger is from contamination by ground-water containing dissolved carbon dioxide from the atmosphere dating from some time after the death of the creature that made the shell. Careful examination of the site in which the shell was found and a curious powdery appearance in shells and pearls so affected by ground water are usually enough to eliminate the danger of false results.

Possibly the most useful material from archaeological sites for dating is charcoal. Professor Libby described it as 'virtually uncontaminable,' and then explained how he learnt has to treat it:

Charcoal can be decontaminated by a combination of physical and chemical laundry. First, plant rootlets must be removed with tweezers under a magnifying glass and then a strong alkaline wash should be used, followed by a careful distilled-water rinse to remove a large number of contaminants, of which the humic acids are a major part. This is followed (actually the order may be reversed and it is not clear which order is preferable) by a treatment with hot hydrochloric-acid solution to remove carbonates, followed by a very careful distilled-water wash and drying in clear, dust-free air.[2]

Although charcoal is from many such points of view an ideal material for radiocarbon dating, there is one great problem with it. How long before the wood was burnt did it die? This is an especially acute problem in virtually treeless areas such as Egypt and the Middle East. Large balks of timber might be used many times over in various different successive buildings on the same site. The radiocarbon date will be the date on which the tree was felled – the day of death. But the date of death could be several hundred years removed from the date on which the palace was sacked and burnt by barbarian invaders, and it is the date on which the wood was burnt that interests the archaeologist. The same sort of problem arises in the case of dating unburnt wood and possibly leather, too, in some cases. 'Short-lived' materials such as reeds or scraps of linen which must have been used within a few years of the date of their organic death do not present this problem, of course, but they are, in general, far less likely to be preserved.

This is a problem for the archaeologist rather than for the radiocarbon scientist, and it can be solved by correct archaeological procedures at the dig. In those first twenty years after the invention of radiocarbon, this problem was recognized along with others which demanded the introduction of important modifications to the procedure in the search for greater accuracy and consistency between laboratories.

Three of these modifications are significant. First, there is the allowance for carbon 13, which is an isotope of carbon 12, even rarer than carbon 14. At one time there was worry that any one of the various processes involved in preparing a sample for dating might change the ratio of carbon 13 to carbon 12 by some unknown selective process which would favour one isotope rather than the other. These fears have been largely laid to rest and it is now usual simply to apply a standard correction factor for the small amount of carbon 13 which can be expected to distort slightly the important carbon 12/carbon 14 ratio, which is the measurement essential for accurate dating.

Secondly, there was the recalculation of the half-life of carbon 14 – i.e. the time in which half the original radioactivity might be expected to disappear. For most purposes it is reasonable to take the half-life of carbon 14 as about 5,600 years. Libby's first accurate experiments and measurements gave a half-life of 5,568 years, and this figure was used for several years. Later work, however, has shown that 5,730 years is a more accurate figure for the half-life of carbon 14, and this is the figure used nowadays by many laboratories, though the officially recommended figure of 5,568 years, set by the Sixth International Dating Conference in 1965, is either used or quoted in most papers. The difference between the two figures is not really of great significance because a simple conversion factor will give the relatively minor change in date between a figure worked on the basis of one half-life and the appropriate date worked on the basis of the other.

The third important modification in technique was the setting-up of an accepted international standard of 'modern' conditions. 1950 is 'now' or 'the present' in the world of radiocarbon dates. A sample of oxalic acid – a normal carbon-containing substance typical of 'life' – was established in the National Bureau of Standards in the U.S.A., and the proportion of carbon 14 to carbon 12 in this sample is considered to be the equilibrium level from which all variations can be counted. 1950 was chosen as the year for 'now' because the major pollution of the atmosphere by atomic and hydrogen bomb testing had not then begun to bring about any major change in the proportion of carbon 14 to carbon 12 in the atmosphere or in living things.

It was, in fact, exactly twenty years after the discovery and announcement of radiocarbon dating that there occurred the event which can be regarded as the formal acceptance of the new technique by the world of scholarship and science as a whole. This was the Twelfth Nobel Symposium, held in Sweden in 1969, officially titled 'Radiocarbon Variations and Absolute Chronology'. It was a meeting of archaeologists, carbon-dating specialists, dendrochronologists and the first of the rising generation of scientists who were deliberately setting out to provide archaeology with new techniques drawn from the sciences.

There had been many international conferences before this on the subject of radiocarbon dating, and not every archaeologist accepts the technique and its results even to this day. Nor did the Nobel Symposium pass any official resolutions declaring that radiocarbon dating was now accepted. But it was at this meeting in Sweden that three American dating laboratories showed that their results had a good average agreement for radiocarbon dates calibrated by tree-ring dating and therefore that an independent calibration for radiocarbon results existed. It was at this meeting that the clock was reset and the variations in the absolute accuracy of the clock were demonstrated. It was shown how radiocarbon readings could be conveniently corrected, and a programme for refining the accuracy still further by international agreement and co-operation was set in motion.

The necessity for resetting the clock and checking it by some external, independent system had become steadily clearer throughout the late 1960s. It had come from getting a larger number of radiocarbon readings on ancient Egyptian materials. It will be remembered that a few specimens of ancient Egyptian material had been used by Libby to show that, in the broadest terms, radiocarbon dating was a viable technique. But as the radiocarbon specialists accumulated more results on Egyptian material, it became disturbingly clear that radiocarbon dates were very often at least two hundred years away from the truth, and sometimes much more. The Uppsala Conference set in motion a programme of special excavations in Egypt to provide samples which could be dated by both radiocarbon and historical methods. In some famous cases where samples of linen from very early First Dynasty tombs were tested in Libby's own University of California laboratories the discrepancy was as great as eight hundred years. All the opponents of radiocarbon dating – whatever the period in which they specialized, whatever the area of the world in which they worked – could choose at least to ignore radiocarbon dates if they wished.

Even from the strictly scientific point of view there were obvious weaknesses in the theory of radiocarbon dating. The radiocarbon method was based on several assumptions, none of which could be proved to hold true in those centuries to which they were being applied – it was essentially an extrapolation into the past of conditions and trends which we can measure today.

The four assumptions upon which radiocarbon dating are based are worth repeating. They are: (1) that the carbon 14/carbon 12 ratio in the atmosphere remains the same, in equilibrium, throughout historic time; (2) that a living organism at any time will have the specific radioactivity caused by carbon 14 which is equal to the specific activity of the atmosphere; (3) that after death the ratio of carbon 14 to carbon 12 can only be changed by the radioactive decay of carbon 14; (4) that the activity of the sample decays at a known rate at all times (exponential decay in fact is specified by the scientists).

In terms of the radiocarbon clock, the first two of these assumptions concern the setting of the clock, while the second two concern the running, or timekeeping, of the clock. No one really doubted that the clock mechanism was there – it was the accuracy of its setting or timekeeping that was in question. Plainly, from the success achieved by the radiocarbon clock in establishing the dates of the broad outlines of the Neolithic Revolution, it could be used to draw out the main lines of human development in broad sweeps. Plainly too, it could be used to establish which of two objects or sites came first and which second, or, perhaps more important, it could establish rough contemporaneity between two objects from widely distant places. But in the only case in which it could be checked in detail against a long and fairly accurate historical chronology, in the case of Egypt, the clock did not seem to be accurate at all; accuracy to 10 per cent was the very best that could be hoped for, and a clock that is between six and ten minutes wrong in every hour is not going to help catch many planes or trains.

In addition to the major problem of the Egyptian dates there was also the disturbing set of variations in carbon dates shown by de Vries in 1958 and 1959. He had taken objects of known age from the last four hundred years and radiocarbon dated them. He had shown that the radiocarbon method did not always give accurate dates over this period of time, though the inaccuracy was much less than that given by Egyptian specimens. Part of the problem raised by 'the de Vries effect' was solved by allowing for carbon 13, as mentioned above, and the remainder of the variations from accuracy detected by de Vries looked as though they might be accounted for by some slight but regular variation in the carbon 14 level in the atmosphere. It looked, for instance, as though there might be a variation dependent on the sunspot cycles of roughly eleven years and some longer underlying, but regular, variation as well. The problems raised by de Vries now seem, with the benefit of hindsight, to be more important for the hints they gave about overcoming bigger problems than for themselves.

It was this combination of uncertainty on the scientific side caused by the work of de Vries, and some other indications, with the disturbing inaccuracy of radiocarbon results compared with historical Egyptian dates that spurred the search for an independent check on the radiocarbon clock. It was natural to turn to the only other known 'clock that runs backwards', tree-ring dating. But we have seen that the only 'absolute chronology', the only set of tree-ring years anchored to our modern dating system, had been obtained in the south-western United States and ran back only to about the start of the present era. The oldest of the Hopi and Pueblo ruins had been studied and it was not clear how dendrochronology could check on radiocarbon dates any further back than that.

In fact the discoveries that provided an independent check on the radiocarbon clock had already been made in the middle 1950s, but work on this apparently very minor phenomenon had been stopped with the death of the man most concerned. The work was taken up again in 1961 when the need for the clock-check started to become apparent and by the mid 1960s – notably at the Monaco Radiocarbon dating conference – it was becoming clear that such an independent check would emerge.

The landmark quality of the Nobel Symposium of 1969 was largely due to the formal statement that an independent check stretching back a full seven thousand years from the present had become available. The first significantly large set of checking figures became available at the Symposium from the American laboratories of Arizona, Philadelphia and La Jolla.

The independent check was a strange inhabitant of the south-western States of the U.S.A. called the bristlecone pine. *Pinus aristata Engelm'* is its official botanical name, but its importance lies in the fact that the oldest bristlecone pines are the oldest living things in the world. At least three living bristlecone pines have been found that are more than 4,000 years old and there are many examples known of trees that are more than 3,000 years old. The present record for longevity is a 4,900-year-old tree still growing in the mountains of the Snake Range in Nevada.

The bristlecone pine grows high up in dry mountains. It is common at 10,000 feet above sea-level and it enjoys places such as the White Mountains of eastern California, where the Sierra Nevada takes the brunt of the rainstorms coming in from the Pacific, so that the White Mountains are in a 'rain shadow'. Because of the dryness the bristlecone pine is slow-growing and its annual rings give a clear account of the climatic changes that have occurred in its long life. Again, the dryness of the area it favours means that there is little undergrowth and even less natural leaf litter on the ground, so there are very few serious forest fires which could destroy old trees or burn away the remains of even older dead trees. Further, because of the dryness, the wood is very resinous and compact and so, alive or dead, it resists the attacks of moisture and decay. To combat the dryness of the climate the tree can preserve and retain its needles for twenty or thirty years and thus survive many years of continuing drought conditions. There is one other general remark that must be made about the bristlecone pine; it has a quite extraordinary beauty of its own, a beauty of fantastic sculptured shapes worked by the wind and rain, the exposure and erosion, of thousands of years of life.

The discovery of the great age of the bristlecone pine was made by Edmund Schulman, in 1954 and 1955. He was working under Douglass, the founder of dendrochronology, at the Laboratory of Tree-Ring Research at the University of Arizona. In his research he was attempting to record climatic change, and trying to discover the underlying role of the sunspot cycle in such change, by studying tree-rings. In 1953 he changed the strategy of his fieldwork to study trees which grew towards the upper edges of the forest zones, believing that he might find them more sensitive indicators of climatic changes than those that grew lower down. In the dramatically named Trail Canyon, near Sun Valley in Idaho, in the summer of 1953 Schulman found a limber pine which turned out to have an age of nearly 1,700 years, and which provided a most sensitive indicator of the climatic changes throughout its life in the variations of its annual rings. In the following two years Schulman extended his searches much more widely throughout the south-west of the U.S.A. and found the really old bristlecones, one in Nevada, the others in the White Mountains on the borders of California and Nevada.

This extraordinary discovery of living objects three and four thousand years old was published widely in the *National Geographic Magazine* in 1958. Schulman died within a very short time of this publication, unaware that he had, by chance, created a tourist attraction of the first order; an attraction that was, in time, to threaten the scientific work which was his main interest. It was, furthermore, another three years before the Laboratory of Tree-Ring Research could obtain further financial support for Schulman's colleague, Charles Wesley Ferguson, to carry forward the work on the bristlecone pine. So it was only in 1961 that Ferguson was able to start gathering more material, extracting core samples from the bristlecone pines. It was another two years later before Ferguson was able to start trying to extend the chronology of the bristlecone pines even further backwards than the great age of the living trees. He then realized that some of the dead and eroded tree stumps, and even some of the stray bits of wood lying about on the ground, in the groves of bristlecone pine, might enable him to build up the longest accurate chronology that anyone had ever dreamed of.

For the next five years Ferguson was engaged on the enormous task of building up this 'master chronology' – finding corresponding patterns of tree-ring growth in all his different specimens; building up sequences of many hundreds of years, and then trying one sequence in with another; eventually tying the whole into our known calendar of years through the living specimens of these extraordinary trees. By 1967 the chronology went back to 4732 B.C. and he had one wonderful specimen containing a 625-year sequence of rings which, according to a radiocarbon date, was at least earlier than anything he had yet built into the chronology. In July 1967 he was able to find another example of this same tree which also included 300 more rings that were not in the original 625. These 300 extra rings were more 'modern' than the 625 and luckily they turned out to give an excellent 'cross-match' with the earliest rings in the known chronology. So Ferguson was able to put his 625-ring specimen in its proper place in the annual sequence and the chronology was pushed back to 5150 B.C.

This chronology covering 7,104 years back from the present was extended by another specimen in the early summer of 1969. The outer 433 years of this new specimen cross-dated with the master chronology so the inner 380 years extended the chronology to a total of 7,484 years, that is to just beyond 5500 B.C. This was the figure that Ferguson was able to present to the Nobel Symposium in Sweden, but even then he had a specimen of bristlecone-pine wood which radiocarbon dating showed to be 9,000 years old, though it could not yet be added to the master chronology because there were no cross-matching pieces to connect it to the 7,484 level. (The work of building a master chronology has gone on and at the time of writing it is understood that the chronology covers nearly 8,200 years.)

In quite another sense this bristlecone pine chronology was 'long'. Even by the end of 1968 the master 'plot' of the tree-ring sequences was 14.2 metres in length at a scale of two millimetres per year (nearly 45 feet long). New specimens were at least approximately radiocarbon dated to find roughly where along this roll to look for a cross-match.

Even the fiercest and most antagonistic critics of the bristlecone pine chronology do not suggest that it can be more than 5 per cent inaccurate at its extremes of coverage. Most scientists accept that it is much more accurate than this, and Professor Ferguson and his colleagues have been able to cross-check their work with chronologies worked out on other long-lived trees such as the giant sequoia and the limber pine, which take them back to at least 1250 B.C.

Some idea of the problems of working in the field of bristlecone pine research was given to the Nobel Symposium, not by Professor Ferguson, but by Professor Libby:

We are making a concerted effort, working with the U.S. Forest Service, the Park Service, and the Bureau of Land Management, to protect these ancient forests and particularly the pieces of ancient wood lying about on the ground, which, as Professor Ferguson has indicated, are in many respects more valuable than the living trees. It is a difficult problem, because we are unable to tell by just looking at a piece of wood whether it is very ancient or not. Our first thought and plan were to pick up all of this dead wood and put it in a warehouse. This was immediately ruled out, however, as being impractical, so now we are in the process essentially of working with the Forest Service, the Bureau of Land Management and the National Park Service, to keep these things under surveillance and to do everything sensible to preserve them. In the meantime we are trying to devise some quick and easy way of telling roughly the antiquity of ancient wood lying on the ground. Professor Ferguson tells a story about finding the oldest piece ever found right by a trail. I guess that is about the state of the art at the moment – largely a matter of luck – but thousands of tons of ancient wood are lying there and we badly need to go through these areas and somehow segregate and protect them.[3]

Even before this, Professor Ferguson had expressed his worry about the problems raised by the tourists:

. . . There is a long tradition (considered by many a moral right) that visitors to the forest, desert and seashore may collect ornamental wood, and the general multiple-use policy of the Forest Service does not discourage such collecting in any except designated areas. The beautiful sculpturing of the bristlecone pine wood makes it particularly desirable, and even in areas where collecting is restricted, wood is still disappearing. In our research in the White Mountains and other bristlecone pine areas we have noted that the more accessible an area is, the less wood there is on the ground and the more saw cuts there are on the trees.[4]

The problems raised by the wood-collecting of American tourists did not seem to perturb the savants of the radiocarbon-dating world unduly. They were more interested to hear how the 7,500-year chronology established for bristlecone-pine wood linked with radiocarbon results.

Professor Ferguson had provided a large number of samples of bristlecone-pine wood from his collection, and each sample had been accurately dated against the master chronology. The samples went to three independent radiocarbon laboratories: the nearby University of Arizona Laboratory, where Professor Paul Damon was in charge, the University of California laboratory at San Diego under Professor Hans E. Suess, and the MASCA (Museum Applied Science Center for Archeology) Laboratory of the University of Pennsylvania in Philadelphia, from which Miss Elizabeth K. Ralph reported to the Symposium.

Broadly speaking, the results from all these three laboratories were in agreement, and though Professor Suess did not allow his findings to be officially included in the final calculation, the chief result of the Nobel Symposium was the production of a possible 'calibration curve' for radiocarbon datings by which all subsequent radiocarbon figures could be accurately related to calendar years back to about 5000 B.C.

This calibration curve is a wavy line throughout its length and it is compared with the straight line of calendar years (as counted basically on the tree-rings of the bristlecone pine). Some of the waviness of the radiocarbon dates may be due to minor features such as sunspot activity, some may be due to inaccuracies and inconsistencies in the counting techniques of the laboratories – these are still, three or four years after the Symposium, matters of controversy among the professionals. But the essence of the matter is that radiocarbon dating gives accurate results back to about 1000 B.C. After that the results from radiocarbon dating get steadily 'younger' until at about 5000 B.C. radiocarbon is giving a date about eight hundred years too young. At an intermediate stage, a radiocarbon date of 2000 B.C. would be a real calendar date of 2500 B.C.

The bristlecone pine chronology, in fact, has saved the usefulness of radiocarbon dating for detailed historical and prehistorical work at a period which is one of the most exciting in the whole development of man, the period when man, after tens of thousands of years of living by hunting and gathering wild foods, suddenly started on the rapid development into civilization. For any radiocarbon date from the period from 1000 B.C. back to 5000 B.C. can now be corrected to the calendar date by simple use of the bristlecone pine calibration graph (see page 337).

The bristlecone pine chronology applies strictly only to a very limited area of the south-western U.S.A., but all the evidence supports the idea that the carbon 14 content of the earth's atmosphere and biosphere is the same at all places at any one time. The carbon 14 content of the dead bristlecone-pine wood at the level of any tree-ring will be the same therefore as the carbon 14 content of any object which was alive anywhere in the world at that same year as the tree-ring was being formed. The simple, closely-confined 'clock' of the ancient trees of the Nevada mountains corrects the time of the universal 'clock' of radiocarbon, whether the carbon is in a Central American maize-cob, an Egyptian papyrus-scroll, a leather sandal from a Danish bog or a piece of charcoal from a prehistoric Chinese village.

The evidence of the bristlecone pine chronology confirmed the evidence of those archaeologists who specialize in Egyptian antiquities, that radiocarbon dating gave a reading some hundreds of years too modern for materials from 1000 to 3000 B.C. The corrections applied by the bristlecone pine chronology agreed very precisely with the corrections demanded by the Egyptologists. In one way this gave a third line of confirmation to the new readings. But more important, it means that radiocarbon dating can now once again be of value to the Egyptologist, since he can now use the technique to settle dates of objects and sites that cannot be decided by the traditional techniques of stylistic comparisons or documentary evidence. Of course, carbon 14 dating can now be used anywhere in the world to give accurate dates of objects and artefacts, as long as the bristlecone pine calibration curve is used to correct dates before 1000 B.C.

But why should the carbon 14 clock have suddenly veered off the true time-line at 1000 B.C. and earlier? The fact that the radiocarbon clock runs fairly accurately for the three thousand years immediately preceding the present implies that there is no fault in those assumptions relating to the rate of running of the clock, to the rate of decay by radioactivity of the carbon 14. The weight of opinion is that the error is in the assumption that the ratio of carbon 14 to carbon 12 has always been the same. If present-day measurements of the radiocarbon remaining in objects which died in, say, 2500 B.C. give a date of 2000 B.C. then there is 'too much' carbon 14 left undecayed – perhaps it is that there was 'too much' carbon 14 in the object originally in 2500 B.C. This is now generally accepted as being the case, but that still leaves the question open as to why there was more carbon 14 in the atmosphere and biosphere.

The Czech specialists at the Nobel Symposium, E. Neustupny and V. Bucha, put forward the explanation that variations in the earth's magnetic field (which is known to have reversed itself several times in the geological history of the earth) allow varying amounts of cosmic rays to penetrate to the earth's atmosphere – and it is, of course, the impact of cosmic rays that forms all the earth's carbon 14. There is virtually no evidence against the idea that the magnetic field of the earth protects against the incoming cosmic rays, and variations on that field should reasonably allow variations in the rate of penetration of cosmic rays, resulting in variations in the rate of production of carbon 14. Strong support for the theory has come from Donald C. Grey of the Isotope Geology Laboratory of the University of Utah.

Grey summarized the position neatly and clearly in a paper which appeared after the Nobel Symposium:

In summary it can be said that short term variations in atmospheric carbon 14 [Author: i.e. the minor wiggles in the radiocarbon-date graph] are explained fairly well by the solar-wind modulation of the cosmic ray flux, whereas the very long-term variations may be largely due to a variation of the geomagnetic shielding of the Earth.[5]

Another interesting question left wide open is whether the variation in the amount of radiocarbon was greater still before 5000 B.C., the furthest date to which bristlecone pine chronology can take us back. There are hints, no more, that is about 8000 or 10,000 B.C. the radiocarbon dates start to come back towards the true calendar dates. This of course implies that the amount of carbon 14 in the atmosphere was more like our present-day levels. That in turn implies something about the earth's magnetic field. In fact these studies of carbon 14 content of the atmosphere and oceans, primarily aimed at strengthening the dating powers of radiocarbon techniques for archaeological purposes, have now opened a new way of studying major changes in the earth's climate and in the interactions of oceans, air-masses and plant-life. The Eighth International Conference on Radiocarbon Dating was held in New Zealand in 1972, and one of the main features of the conference was the production of standard corrections which any archaeologist could apply to radiocarbon dates, to bring the dates into line with tree-ring chronology and therefore with the normal count of human years. A considerable amount of work had been put into producing these new standard calibrations of the radiocarbon results by getting a great deal more information from radiocarbon dating and tree-ring counting of specimens of both the bristlecone pine and the giant sequoia. Two teams had been involved, Dr H. N. Michael and Dr Elizabeth Ralph from the University of Pennsylvania, and Drs P. E. Damon, A. Long and E. I. Wallick from the University of Arizona. There was great disappointment when it was found that the corrections advised by the two teams did not agree with each other. The conference could not recommend any official calibration and correction of radiocarbon dates. Dr Suess, of the radiocarbon dating laboratory at La Jolla in California, has still refused to publish all his results (which were the basis of the first calibration table produced at Stockholm) in tabular form.

There is a basic disagreement here. Suess believes that all the wriggles and curves shown up in his original calibration graph may be genuine products of disturbances in nature – irregularities in the production of carbon 14. The other two teams have used different methods of smoothing out the kinks by mathematical averages over lengthy sets of readings so as to get useful correction figures which archaeologists not expert in the subtleties of radiocarbon dating can use in practical applications. Their curves also have the advantage that any one radiocarbon result can only give one date for the object examined whereas the kinks and wriggles of the Suess line give places where two alternative dates can be given for one radiocarbon result.

Very recently, however, Dr V. R. Switsur of the Radiocarbon Dating Research Laboratory at Cambridge University has compiled a table of figures showing the corrections demanded by the Arizona and Pennsylvania calibrations back to nearly 5000 B.C. He has then taken the mean of the two corrections and he proposes that this mean correction should be used as a standard recalibration of radiocarbon results. He points out that the difference between the two corrections 'in most cases is comfortingly small and probably within the limits of the standard deviation provided with the radiocarbon date. Only during the radiocarbon time period 3900 bp to 5500 bp does the difference exceed 100 years, the average difference for the whole of the time covered being only 34 years.'

The acceptance of the mean between the two corrections as the practical way out of the difficulty is described by Switsur 'as the best corrections available at the present state of the art'. He adds, 'The conversion factors are based on about 600 individual measurements and it will be many years before a sufficiently greater number of measurements will be made to alter the shape of the deviation curve significantly. Hence there is little likelihood that correction factors that are very different from these given will be derived.' Switsur's proposals were published in the June 1973 edition of *Antiquity*, and were immediately criticized by Richard Burleigh, of the British Museum, in the next issue of the same journal. Elizabeth Ralph of the University of Pennsylvania in the August 1973 issue of the *Masca Newsletter* publishes her full figures of radiocarbon dates and dendrochronological dates, and seems to be nearer to Burleigh's position than to Switsur's.

As an outsider, it seems possible to sum up the position thus – the present systems of calibrating radiocarbon dates by corrections based upon the tree-ring data are quite satisfactory for all reasonable archaeological purposes, and they will give a date which is usually within 1 per cent of historical accuracy, but the archaeologist must realize that the scientists have not yet reached a definitive or accepted solution to the problem in their own minds and they are observing scientific caution in warning that there may be further, though slight, changes still to be made in the calibration. The archaeologist is therefore warned that when he publishes the radiocarbon dates on his findings he should publish a note saying just how the calibration of the basic radiocarbon data has been performed, so that later generations can correct the historical dates if and when the calibration corrections are revised.

The argument among the scientists has become very detailed by now. Dr Ralph for instance finds that her figures show a steady variation with a period of nine thousand years, and a possible variation with a period of four hundred years. These would be caused by geophysical factors such as variations of the earth's magnetic field or even variations in the sun's magnetic activity. Burleigh is interested in a long-term dendrochronology being built up on samples of Irish bog-oak by A. G. Smith and his team at Queen's University, Belfast, which may eventually span a period of eight thousand years. This should provide a further independent check on the bristlecone pine, and should reveal whether there is any peculiarity in the climatic conditions of the south-western states of the U.S.A. which may have caused the radiocarbon content of the bristlecone-pine wood to vary from Libby's original assumptions.

But for the practical archaeologist the radiocarbon clock – the clock that runs backwards – has been checked, reset and calibrated so that it is now available for anyone to use with confidence, as long as due allowance is made for the scientists' reservations.

Even before these latest adjustments had been made, however, the recalibration based on the tree-ring figures had produced a great many important recalculations of the earliest radiocarbon dates. In the case of the European Neolithic this made many objects and sites at least two hundred or even five hundred years older. This turned the prehistory of much of western Europe upside-down. The intellectual upheaval caused by this shattering of earlier and long-held views has been the seed-bed from which the 'new archaeology' has sprung in Europe. (See Chapter 15.)

1. E. Neustupny at the Stockholm Nobel Symposium, 1969.
2. W. F. Libby at the Stockholm Nobel Symposium, 1969.
3. ibid.
4. C. W. Ferguson, 'Bristlecone Pine: Science and Esthetics'.
5. D. Grey, *Journal of Geophysical Research.*

6

A Modern Radiocarbon Laboratory

The actual production of radiocarbon dates has, up to now, been concentrated at museums or in university laboratories. It has been those interested in radiocarbon dating for professional reasons, either the archaeologists attached to museums or the physical chemists in the universities, who have done the work. Since radiocarbon dating was not originally the prime object of their professional careers, the work has often been done in buildings which were not purpose-made for dating or radioactivity work. Furthermore, those who have been doing the work have had, perforce, to help their colleagues and provide radiocarbon dates for those excavators and archaeologists who have not been able to obtain their own equipment.

But now we are beginning to see the first radiocarbon-date factories – special units or laboratories set up with the sole purpose of providing radiocarbon dates for any scholar or scientist who may wish to have his material examined. There are two reasons for this development: first, it is an attempt to make radiocarbon dating a cheaper process, by applying some of the lessons of production-technology; and secondly it is because so many different lines of research, as well as archaeology, are now 'customers' for a service which can give radiocarbon dates to samples.

It costs about £100,000 (a quarter of a million dollars) to set up and run a 'radiocarbon-dating laboratory' using the most modern techniques. The whole building, its equipment and its staff is essentially devoted to the quite extraordinary task of identifying one carbon 14 atom among every one million million ordinary carbon 12 atoms, which is the proportion of radiocarbon to ordinary carbon in living matter. In practical terms this means that from one gramme of pure carbon they can expect just 13 'counts' in each minute. If the specimen being examined was alive 60,000 years ago (this is the limit to which radiocarbon dating can reach at the moment), this laboratory must find one carbon 14 atom in *one thousand* million million ordinary carbon atoms.

It is usual in scientific papers and writing to refer to the accuracy of such a process – the accuracy of radiocarbon dating – but what the engineers and designers of such a laboratory are really seeking to achieve is discrimination. They must find those thirteen counts in each minute, and they must be certain they come from the specimen they are studying, not from anywhere else. They must therefore be able to discriminate those thirteen counts from all the other disturbances going on in our universe, and they must be able to discriminate between genuine material to be examined and all other material. To achieve this power of discrimination is a matter of design, architecture, engineering, electronics and high-quality process-control. It is these, perhaps less glamorous, facets of technology which enable the archaeologist to remark in his paper that such-and-such a level was dated to 9750 ± 150 B.C.

Let us take one of the first examples of a purpose-built radiocarbon-dating laboratory, the Radiocarbon Laboratory at the Scottish Research Reactor Centre in the new town of East Kilbride, near Glasgow. It was opened in September 1972, and it was, from the start, designed to serve a wide variety of universities and research institutes, and a wide field of different sciences, providing radiocarbon dates for many other people as well as archaeologists.

This laboratory consists, in fact, of three laboratories in sequence. The first, the chemistry laboratory, receives the samples from the field, and the basic task here is what Libby has called 'laundry'. Every scrap of matter which was not part of the original living specimen must be removed. The four most common contaminants are rootlets from modern plants; humic acids (the acid chemicals produced in the living topsoil, or humus, by rotting vegetation and the myriads of micro-organisms that live on it); limestone, which contains carbon, and which may easily be found in any mud or clay surrounding an excavated specimen especially if the limestone is present as small sandlike particles; and finally the chemicals called carbonates many of which are easily soluble in water and which tend to replace, in a chemical reaction, the original material of shells when they become buried in damp ground.

Rather complicated chemical techniques may be needed to remove some of these contaminants, and even when simple techniques are satisfactory they have to be carried out with great care and precision to see that as little as possible of the carbon of the original specimen is carried away with the contaminants. Vast quantities of patience are also required at this stage of the processing, because usually someone has to perform the laborious task of teasing out and picking off all the rootlets that have penetrated the specimen, and this can only be done with needle-point and tweezers.

When the sample has been cleaned it is moved into the 'vacuum line laboratory', where the objective is to get all the carbon atoms out of the original material and to make them into a chemical or substance which can be easily and accurately measured for its radioactivity. The first stage of the process is simple; the sample, whether it is peat or bone or wood or charcoal, is totally burned in an atmosphere of pure oxygen. This puts all the carbon atoms into carbon-dioxide gas, much as Libby had done with his first specimens. This carbon-dioxide gas is fed into the first stage of the high-vacuum system, where it is purified from the other gases generated in the burning, and then it is exposed to molten lithium so that all the carbon atoms from the original sample are incorporated into lithium-carbide. This lithium-carbide is reacted with pure water to produce acetylene, a gas which can then be purified again so that all traces of lithium and other substances have been left behind. The acetylene is then reacted in the presence of a catalyst and converted into benzene, a liquid which consists of little but carbon and hydrogen and which should contain virtually all the carbon from the original sample.

The object of this long series of processes is to concentrate the carbon from the original sample. In the once-living matter the carbon atoms formed only a small proportion of the whole collection of atoms, but in benzene carbon forms a much larger proportion of each molecule, and furthermore benzene is a substance which can easily be handled. The object of performing all these reactions under conditions of high-vacuum is to exclude the atmosphere and thus to avoid contamination of the carbon from the original specimen by ordinary modern carbon dioxide in the air.

But the use of a liquid as the final carbon-containing substance in the counting process is claimed by its adherents as one of the most important advances in technique in the history of radiocarbon dating. It is certainly spreading among major counting laboratories, as opposed to the original practice of doing the counting on a gas or solid form of carbon. The liquid system, for example, concentrates the carbon into a smaller sample. This is easier to handle and enables a smaller, more easily designed counter to be used. The smaller counter, in turn, poses fewer problems in designing systems which cut down the 'background' radiation. It is also claimed that it is much easier to change the samples inside the counter when liquid is used – in the earlier days it could take a whole day's work cleaning out one gas from the counting chamber, purifying the chamber, and putting in another gas sample. The gas-counters suffered from the problem that background radiation is known to vary with atmospheric pressure, and, since no human ingenuity can change atmospheric pressure, there was a built-in uncertainty in the results which does not occur with liquid-counters. Finally the liquid counting systems, which use commercially available liquid scintillation counters in the final stages, can also take 'control' samples. These are samples made up of benzene with a known radioactivity, so that the total count shown by the counter when the control is inside it gives a measurement of the total of radioactive 'events' or 'counts' which are caused by outside sources or background radiation. This figure of unwanted radiation can be subtracted from the count given by one of the ordinary samples so that the true amount of radioactivity in the sample can be accurately measured.

However, before the benzene can be put into the final counter a fluorescent material, based on phosphorus, is added. The radiation emitted by carbon 14 is comparatively weak and travels only a short distance through liquid, but the fluorescent material glows whenever a pulse of radiating energy passes through it, and therefore gives off a flash of light whenever one of the carbon 14 atoms disintegrates. Highly sensitive photomultiplier tubes detect these scintillations, where the human eye could see nothing, and send off an electrical signal, a 'count', for each flash of the fluorescent material.

Liquid scintillation counters can be found in many laboratories. They are especially common in medical and biological centres which use radioactive isotopes for tracing the movements of drugs, hormones, insecticides and other materials, both natural and synthetic, through living systems. The presence of radioactivity in samples is detected by the medical researchers in the same way as the radiocarbon atoms in former living materials are found in the dating laboratory. The archaeologist is thus able to benefit from the very considerable design and engineering effort that commercial companies have put into the development of the liquid scintillation counters, and especially from the industrialists' efforts to cut out all background radiation in their machines.

Nevertheless, the counting laboratory at East Kilbride has been built underground with walls, floor and ceiling constructed from concrete twelve inches thick - made of materials specially selected to have the lowest possible amount of naturally radioactive chemicals in them. In addition to this the first two stages of the laboratory, built above the counting chambers, contain further sections of massive concrete so placed as to give maximum extra protection in the direction of known sources of radiation in other laboratories or industries in the surrounding area. All this is to cut down the number of unwanted radiations and radioactive particles passing through the counting chambers. Only if the background radiation is enormously reduced can the counters achieve the high level of discrimination necessary to count the comparatively rare and weak emanations from the few radioactive carbon atoms.

These expensive precautions taken when the laboratory was being planned and built enable it to run now on something like 'factory-production' lines, without the necessity for constant readjustments and recalibrations. The production line efficiency allows one sample to be measured for radioactivity while the next one is passing through the high-vacuum preparation stages and a third is being cleaned in the chemistry laboratory. The whole of the pre-treatment and processing of a sample normally takes two days, and measurement may take as little as one day, though if the sample is very old and small it may be necessary to leave it in the counter for as many as fifteen days to get a satisfactory number of readings. This new laboratory has been designed and equipped to turn out four hundred dates a year, but the buildings have the capacity to allow for extra equipment to be installed and throughput could be doubled if the need arises.

Only a small proportion of the capacity of the laboratory is being taken up in measuring the ages of traditional archaeological samples, in the sense of samples excavated to throw light on the history and prehistory of man. The greater part of the laboratory's work is concerned with samples which illuminate the history of the environment; typical examples are samples of shells or peat, from beds covered now by more modern deposits, which, when dated by their radioactive carbon content, can be used to reconstruct a history of the movements of the glaciers of the Ice Ages as they advanced from Norway to cover Scotland. Other samples from the Shetland Islands have dated the warmer periods that came between successive advances of the ice. Dates from samples taken from lake beds and peat-layers in the Lake District of England and the north-west of Scotland are being combined with analyses of the pollens found in these deposits to build up a history of the climate in its relation to the vegetation of these areas – an ecological history. Yet other samples are sent to this laboratory by the scientists of the British Antarctic Survey.

These studies of ecological and environmental history seem fairly obvious extrapolations from the study of man's history. It is a big step further to apply carbon dating to modern problems, but the step has been taken in the study of water-resource problems. In many parts of Britain and in areas like the north-eastern United States, where modern industry is making ever-increasing demands on water, there is a real threat of a water shortage in the future. It is customary to tap underground reserves of water by deep bore holes or artesian wells to supplement the flow of surface water in rivers and reservoirs, but in the areas of heaviest demand we are becoming increasingly and uneasily aware that the underground water-levels are steadily dropping as we draw more and more from them. Water supply authorities wish, therefore, to know the rate at which these underground 'aquifers' – layers of water-containing, porous rock – are recharged by water slowly percolating down from the surface. Rainwater contains, in fact, a small amount of radioactive carbon 14 washed out of the atmosphere where it was formed, and when the rainwater eventually reaches the natural underground reservoirs it is out of contact with the atmosphere, so the radioactivity slowly decreases as the carbon 14 atoms disintegrate. This gives a measurement of the amount of time the water has spent underground and hence the rate of recharge of the aquifers can be calculated. But for the radioactive-dating laboratory such studies mean an immense expenditure of time, because the preparation of a worthwhile sample involves recovering all the carbon dissolved in a thousand litres of water.

While these studies of the environment and of ecological history have been made possible only by the development of radiocarbon dating and are therefore derived from the application of laboratory science to archaeology, they will eventually repay their debt to archaeology. Where once it was only possible for the archaeologist and prehistorian to study man in the distant past by examining the remains that man had left behind in the shape of pottery, carvings or artefacts, now it is becoming possible to study the traces of past environments. When we can put the two histories together we should get the more significant picture showing the effect of the environment on man and the effect of man on the environment.

7

More Clocks: The Breakthrough

Radiocarbon dating did far, far more than provide the archaeologist with dates. It is one of the few scientific discoveries that genuinely merits the use of the word 'breakthrough'. It gave a fresh turn to the way men thought about the discovery of the past. It revolutionized the profession of archaeology; it changed the way in which excavations and digs were conducted. And in several cases it turned upside-down our reconstructions of the order of events of the past.

In 1947 Jacquetta and Christopher Hawkes could write:

> Archaeology has now accumulated enough information for it to be rare for experts to have any difficulty in dating finds or assigning them to a recognized culture, merely by looking at them. This certainty with which an archaeologist can date a scrap of bronze or even a wretched potsherd, to some people seems almost miraculous, to others dubious. But it is a matter of experience, of having handled such things for years, together with some visual and tactile sensibility: very much the same qualifications as are required to distinguish the quality of foodstuffs or materials or the work of different painters. A form of connoisseurship though often without the usually accepted implication of elegances. Almost all the things made by man can with experience be recognized this way and be used with more or less accuracy for the identification of the culture to which they belong. This special effluvium of each age down to its smallest items is a fascinating thing.[1]

What happened when radiocarbon dating made its impact on this world of connoisseurship in discriminating between the effluvia of different ages? Twenty years after the Hawkes's book was published Frederick Johnson, curator of the R. S. Peabody

Foundation for Archeology at Andover, Massachusetts, looked back. Twenty years before,

Because chronology is essential to the understanding of prehistory, archaeologists have made great efforts to extract measures of time from their data, which are hardly adequate for the purpose. With few exceptions this extraction was by inference and guessing; nevertheless the various systems presented have been staunchly supported. Libby's provision of a means of counting time – one that promised a definable degree of accuracy and world-wide consistency – caused all sorts of consternation because many of the new findings threw doubt on the validity of some established archaeological opinions.

So right from the start, as far back as the time when Libby asked for, and got, his advisory committee there was controversy. Johnson writes:

There was no way in which archaeologists could check the radiocarbon date on a sample of unknown age except by comparing it with archaeological opinion. When, as at the beginning, there were frequent disagreements concerning variations of from several hundred to more than 1,000 years, some archaeologists believed that their credibility was threatened and they rebelled. At this primary stage, a question of authority arose: Who was right? Some believed that archaeology had the prerogative and that it should declare that its inferences were valid, even though there was a good chance that some of them were wrong. One somewhat typical comment was based in part on this conviction: the author of a very reputable monograph wrote in 1951, 'We stand before the threat of the atom in the form of radiocarbondating. This may be the last chance for oldfashioned, uncontrolled guessing.'

One of the great controversies of American archaeology of the pre-radiocarbon days concerned the Adena and Hopewell cultures, centred round Ohio but covering in one form or another most of the eastern U.S.A. By the 1930s the general consensus was that the Adena culture dated from about A.D. 500 to 900, and that the Hopewell culture was a major religious and ceremonial development from it which dated from A.D. 900 to 1300 in broad terms. But, because of the uncertainties in dating, it was impossible to tell whether the different variants, especially of the Hopewell culture, as it spread outwards from Ohio to Kansas, Illinois and Wisconsin and into the Great Plains, came before or afterwards in time – whether they were signs of 'development and progress' or 'degeneracy and impoverishment'. Then in the early 1940s there was a theory that pushed both cultures backwards in time; but in the following swing of the academic pendulum Hopewell returned to around A.D. 1200. Johnson says of this controversy:

The time and energy spent in trying to establish the age of Hopewell ended in nothing but a succession of frustrating estimates and furious debates. New excavations and publications, together with the expansion of analyses, added fuel to the debate, but the prospects of solving the chronological problem by purely archaeological means did not improve. Radiocarbon dating broke the back of the dilemma dramatically.[2]

At first there was 'confusion worse confounded', because the very earliest set of dates provided by radiocarbon reversed the order of things and made Adena later than Hopewell. But some of the original material used for dating was very doubtful – it had never been collected for carbon dating and the first excavations of the original Hopewell site were done as far back as 1891. So everyone had to take a fresh look at the whole matter, details were reassigned from the classification of one culture to that of another, fresh excavations were undertaken to obtain material suitable for carbon dating. When Johnson wrote in 1967 there were still too few dates to be certain, though it appeared that Adena did predate Hopewell after all, but the whole period had to be put back some seven or eight hundred years – Adena to 800–100 B.C. and Hopewell to A.D. 100 to 500.

The statement that radiocarbon dating caused a 'revolution' in archaeology and archaeological ideas about chronology is confirmed by Johnson, and he also confirms the fact that radiocarbon led to major changes even in the techniques of excavation: 'This return to the trenches for a more careful look at the provenance of the samples – often for the purpose of proving the radiocarbon dates to be erroneous and useless – resulted in refinement of methods of recording in the field, in order to determine more precisely associations of samples with levels.' Some of the later techniques for applying laboratory science to the solution of archaeological problems have compelled the archaeologist to change his field techniques even further.

There were further changes brought about by radiocarbon dating. The penetration by the nuclear physicists and chemists under Libby into the 'humane' field of archaeology led to many others following after them. Indeed it may be that the greatest effect of radiocarbon dating was precisely this – to induce such a change of outlook that men and women engaged on the problems of discovering what happened in the past actually started looking to the scientists in different fields to see if they could provide techniques which would help to solve the problems of research into prehistory. Much later those who would call themselves 'historians' rather than 'prehistorians' learned to look in the same direction. By the same token, people who worked in physics, chemistry or genetics laboratories began to think, sometimes, that they could produce techniques and machines which would help the archaeologist. Radiocarbon dating was the first successful technique of the laboratory that made a really serious impact on archaeology. It was to be followed by many others, and its success also revived interest in techniques such as pollen analysis which had been tried before and to a greater or lesser extent discarded.

One further broad-scale effect of the coming of radiocarbon dating was that it, and many of the other techniques it brought in its train, strengthened and brought to the fore the relationship between the history and prehistory of man on the one hand and what might well be termed the 'history of the environment'. The importance of seeing man in his environment had begun to be appreciated independently within the discipline of archaeology shortly before the coming of radiocarbon dating – we have seen that Zeuner was Professor of Environmental Archaeology and this appointment in Britain reflected a trend which had started in the U.S.A. a few years earlier. But the study of prehistoric environments demanded techniques analogous to those used in the study of contemporary environments, the techniques of botany, physiology, biology and so on. Some of these techniques were comparatively simple – the recognition and analysis of animal bones for instance. But the technologies of these other sciences were developing and these developments spread into archaeology in turn, as the archaeologists started asking questions which only the botanist or biologist could answer.

Naturally enough, once radiocarbon dating had shown that physico-chemical methods could provide archaeology with its first successful clock, the next thing that people looked for was another clock, and preferably a clock that would be applicable where radiocarbon dating could not work. Again naturally enough, a lot of attention was immediately directed towards other techniques using radioactivity, and within a few years at least three more significant and useful radioactive clocks were found. But, unlike Libby's technique, these were more closely related to the earliest methods of determining age from radioactive isotopes, of the type described by Zeuner back in 1946.

The most valuable of the radioactive clocks after radiocarbon has proved to be the 'potassium–argon dating' method. There are two chief reasons for its success. First, there is the fact that potassium is one of the most common elements on earth, in fact it makes up 2.8 per cent of the total weight of our planet and some of it can be found in practically all the known minerals that make up the earth's crust. Secondly, because of the length of the half-life of the isotope involved, potassium 40, the potassium–argon method covers a period in the history of the earth which we particularly want to be able to date – the period from the furthest reach of radiocarbon methods, about 50,000 years ago, back to the eras when time becomes 'geological'. (It should be said, however, that some experts dispute this claim.) This span of time covers the first emergence of man as a separate species and stretches back to the times when 'absolute' dating becomes a rather meaningless term. Undoubtedly the most significant feat of potassium–argon dating so far has been achieved with samples of minerals from the beds of the famous Olduvai Gorge in Tanganyika, where Dr Louis Leakey did more than any other one man to date to show how man became man. Potassium-argon gave the date of Bed I at

Olduvai, where Leakey found Zinjanthropus, popularly known as 'Nutcracker Man', as one and three-quarter million years old – and thus doubled the time-span that scientists had previously allowed for man's existence. The same technique also showed that very primitive stone tools – of the type known as Chellean – must have been more than half a million years old because they were overlaid by volcanic rocks of that age.

'Natural' potassium contains more than 93 per cent of the normal, stable isotope, potassium 39, and nearly 7 per cent of potassium 41. Only 0.0118 per cent is radioactive potassium 40 which has a half-life of more than a thousand million years. It is estimated that when the earth was formed there was about 0.2 per cent of this radioactive potassium 40. The potassium 40, when it does disintegrate, can do two different things – about 90 per cent of it becomes calcium 40 atoms and 11 per cent of it becomes argon 40 atoms. Argon is one of the so-called 'rare' gases that make up a very small proportion of our atmosphere – helium, neon, krypton and xenon are the others. But argon is the most common of the rare gases and it can be measured more easily than most other elements, even in very small quantities, largely because it does not react easily with other elements. Because the half-life of potassium 40 is so long, quite appreciable quantities of argon 40 build up in minerals containing potassium, but there will still be measurable quantities of potassium 40 left in the rocks. The proportions of potassium 40 to argon 40 in any particular piece of mineral can therefore be measured, and a comparatively simple mathematical formula can give the age of the rock containing the mineral as long as the half-life of potassium 40 is known.

This, of course, is an oversimplification of the procedure and the precautions that have to be carried out to obtain a potassium–argon date – allowances have to be made for the possible original content of argon 40 when the rock was first formed; precautions have to be taken against contamination of the sample by argon in the atmosphere; allowances have to be made for the other isotopes of argon which may be present; only certain types of mineral are suitable for carrying out the dating process. It is even more important, if that is possible at all, for the archaeologist to have carried out his digging and stratigraphic recording so accurately that he can definitely pin the sample of mineral into the level he wishes to date. But the method has been so much improved and the accuracy has become so great that potassium–argon dating 'has established itself as of special importance'.

It has proved possible, by taking large numbers of samples of minerals, and applying statistical analysis to improve on the accuracy achieved by laboratory methods, to date certain samples that are within the extreme limit of the time-range of radiocarbon dating – about 70,000 years back. This has established a link between the two chronologies, though 'absolute' dates at this distance back in time are accurate only to the extent of some thousands of years.

This development of the potassium–argon dating method, which is now used extensively in plotting the early history of man and the very recent geological history of the earth, must be seen, however, in a broader perspective than that of archaeology alone. It was quite soon after the first discovery of radioactivity that Rutherford showed that dating of geological events by some method of this sort would be possible – he demonstrated that a crystal of a mineral called 'fergusonite' must have been formed more than 500 million years previously because it contained 7 per cent of uranium and 1.8 cubic centimetres of helium – the helium being formed from the alpha particles emitted during the radioactive disintegrations of uranium, as explained in Chapter 2. As long ago as 1937 the presence of argon 40 atoms in the atmosphere was demonstrated as being due to the disintegration of radioactive potassium 40. But it is only the development of extremely precise techniques of measurement such as flame-photometry and mass-spectrometry, the latter requiring large and expensive machines, which has enabled the small quantities of potassium and argon involved to be determined with sufficient accuracy to provide dates. These techniques were not developed primarily for archaeological purposes – indeed there is no reason to believe that those who developed them had any idea that they would ever be used to date bits of rock spewed out from East African volcanoes whose eruptions were observed by the creatures who had ceased to be apes but had not yet become men. In pressing the potassium-argon dating technique to its limits recent practitioners have been using even more esoteric methods such as isotope dilution analysis and neutron activation analysis.

Two further clocks based on radioactive disintegration have also emerged as useful. But they are both limited in application as far as archaeology is concerned. There is the rubidium–strontium dating method, which is largely used by geologists to determine the age of older rocks, and which has been used to date samples of moon rocks. It works on exactly the same principle as potassium–argon dating and this is also true of the dating system based on the radioactive decay of protactinium 231 atoms into atoms of thorium 230. This latter method has been used to give 'absolute' dates to the contents of deep-sea 'cores'. These 'cores' are the long samples obtained when a hollow tube is drilled down through sediments of the sea bottom. The sediments are largely made up of the shells of the minute sea creatures called *Foraminifera* which slowly sink to the bottom of the oceans after the end of the lives of their inhabitants. They carry, therefore, a record of the life that went on near the surface, where such changes as the variations of climates are recorded in the differing chemical composition of their shells. The cores are collected by research ships and brought back to laboratories such as the Lamont Geological Observatory the Scripps Institute of Oceanography at La Jolla, California, or the Swedish Deep-sea Expeditions. By comparing the variations in shell constituents representing colder or warmer climates with the time-scale given by the protactinium–thorium method the broad sequences of warmer and colder spells can be built up to show remarkably good agreement with those built up from observations of the geological record on land.

One major gap exists in this system of clocks depending on radioactive decay. The existing systems link up in their time-spans to give virtually complete coverage from the present day back to the formation of the planet – but none can measure the age of fossil bones directly. Radiocarbon dating can provide the age of bones not yet fossilized, so long as there is some remnant of the once-living organic material of their original composition. But when the carbon-containing material has disappeared and been replaced by minerals from chemical reactions with the rocks, dating proves impossible. The potassium–argon method has been tried, but has so far failed to achieve a dating on fossils.

Bone is, along with pottery, the commonest feature of the archaeological record – we all know that excavators dig up bones. Bone in its fossilized form provides our knowledge about the life forms of the earth over the millions of years of geological time, and non-fossilized bones that show human burial customs and eating habits are one of our main sources of information about men of the past. If we could find a method of dating bone absolutely this would be the best of all geological clocks. Naturally men's minds were turning to the possibility of finding a clock in bone at the time when Libby was developing his radiocarbon clock. However, in the early days of radiocarbon dating the technique could not be applied satisfactorily to any but fairly modern specimens of bone. The ability to date even non-fossilized bone by radiocarbon techniques is a very recent refinement which has been developed largely by one of Libby's early colleagues, Rainer Berger.

We have already seen how the radioactive dating of rocks had been seen as a possibility ever since the days of Rutherford; yet it has only been developed as a practical technique subsequent to the discovery and establishment of radiocarbon dating. This demonstrates the importance of Libby's work in achieving the 'psychological breakthrough'. Once he had shown that a clock that runs backward could exist, then other scientists soon found other clocks using mostly knowledge that already existed. Zeuner's work in the late 1940s (see Chapter 2) had also impressed on the archaeologists the importance of finding dating mechanisms. Very few scientific developments of importance occur in a vacuum; the desired objectives in any developing field are usually widely known and the final 'great discovery' is often simply a question of who gets there first and what line of approach they have used.

So it is not surprising that, while Libby was working in America, there was a different approach to finding a clock-mechanism being developed simultaneously in Britain. This was the attempt to date bone by chemical analysis. It did not produce a universal and absolute clock (as Libby did) but it produced a mechanism, the fluorine test, which gave a very accurate relative-dating technique, which in any one locality could show, also, whether a specimen was truly contemporary with the level at which it was found.

The great triumph of the fluorine test was its part in revealing that the 'Piltdown Skull' was a fraud, and this episode will be described in Chapter 13, but the wider importance of the fluorine test for archaeology lies in its power to establish whether or not a specimen is contemporary with the level in which it is found. There are two important mechanisms which can place a piece of archaeological evidence in a layer of earth to which it does not truly belong, and which can therefore mislead the archaeologist about the chronology of the site. Human burial practices are the first and most frequent source of confusion. A body decently interred below ground is by definition placed in a 'layer' or stratum of earth in which it does not chronologically belong. It is an 'intrusion' into the record of an earlier time. For the archaeologist this problem is particularly acute in those cultures and societies where the dead are normally buried beneath the floor of their own family's living-site. It can also be a major problem in any site which has been inhabited more or less continuously for long periods, such as in many of the great Middle East 'tells'. The other common mechanism placing evidence in the wrong time-horizon is the action of water in rivers or streams. The remains of a camp site or of a whole village on the bank of a stream can easily be washed away and deposited, perhaps many miles downstream on a spit of sand or gravel which consists of material having no demonstrable chronological relationship to the source of the archaeological evidence. The fluorine test, along with other similar chemical tests, can show the archaeologist whether the objects he has dug up are truly contemporary with the layer in which he has found them, or are intrusions from another age.

The idea of using chemical tests of bone to settle this sort of question is quite an old one. Something similar was used to show that the 'Calaveras' skull, found in a Californian gold-mine digging in the last century, was neither a murdered contemporary miner nor a prehistoric man. The French scientist Carnot had actually proposed the use of fluorine analysis for this sort of work before 1900. But it was not till the end of the Second World War that the technique was finally developed. We call it the fluorine test but it is really a battery of tests for determining whether a wide variety of chemical constituents of the buried bone are present in proportions consistent with the presence of the bone in that particular soil for a specified period of time.

The processes that affect bone once it has been buried are many and complex. The fatty material disappears quite rapidly, as might be expected, but one of the chief organic components of bone, the protein called collagen, takes much longer to disintegrate, and the rate at which it disappears depends entirely on the chemical nature of the environment in which it is buried. Where the soil is permanently frozen or where air and bacteria cannot get at the bone the organic collagen may not be destroyed for tens of thousands of years; yet on the other hand different soil conditions may destroy bone collagen very rapidly. The preserved mammoths of Siberia show one extreme, though a more important example is the discovery of some traces of organic matter in the bones of fossil fishes 300 million years old found in the Devonian shales of Ohio. At the other extreme there is the case of what was thought to be a 'fossil horse' dug up in Gibraltar late in the last century; the discovery caused such excitement that further excavations were carried out and the feet of the animal were found near by with the metal horseshoes still in position, for it turned out to be a military 'charger' buried only some twenty-five years before it was excavated.

If any considerable amount of organic material remains the bones can be dated by the radiocarbon method. But neither radiocarbon nor potassium–argon can date fossilized bones. Fossilization consists of two processes – the filling of the pores of the bone with new mineral matter such as iron-oxide or lime; and the gradual alteration of the mineral, part of the bone by atoms of elements in the soil taking the place of atoms laid down by the once-living body. The mineral part of living bone is called hydroxy-apatite and the chemical change involved when a bone is fossilized is very often the replacement of the hydroxyl in hydroxy-apatite by fluorine and uranium.

Fluorine can almost always be found in the water in soils all over the world, although the proportion of fluorine in these 'ground-waters' naturally varies from place to place depending upon the underlying geology of the area. Uranium, however, is less common in ground-waters, but its presence may be extremely important in certain cases.

Plainly the amount of fluorine that is incorporated in a buried bone will increase progressively with time, and in this sense a 'clock' mechanism is set up. This clock is, however, so dependent on the local chemical conditions and constituents of soil and water that it only gives a local time. It cannot be used all over the world, but in any one place it provides an accurate and easy-to-apply check on whether two bones found close together have been in that soil for the same length of time. A famous example of this technique in action concerned the 'Galley Hill Skeleton'. This was discovered in 1888 in river gravels by the Thames alongside very early Stone Age hand-axes and the remains of an extinct elephant which was known to date from before the time of Neanderthal Man. Much later, in the 1930s, this skeleton became very important for the Swanscombe skull was then found at a considerably lower level, and the Swanscombe skull clearly was that of a very early type of man. The fluorine test showed conclusively, as early as 1948, that the Swanscombe skull was truly very old, but that the Galley Hill skeleton was not of the same age as the gravels in which it lay or as the elephant remains near it. It was not until 1959 that it was decided that enough of the Galley Hill skeleton could be spared to allow a radiocarbon date to be given, and this showed that Galley Hill man was just over three thousand years old. The skeleton had therefore been buried in Bronze Age times.

Measurement of the amount of nitrogen in a bone gives a reading of the amount of organic material left undestroyed, and this measurement is often used in conjunction with the fluorine test to give a cross-check – because two bones buried at the same time in the same place should not only have taken up the same amount of fluorine as each other, but should also have lost the same amount of organic matter. Uranium uptake, where it occurs, will also give a clock effect which is purely local. Uranium contents of a bone can be used just as fluorine contents as a check on contemporaneity of two specimens, but with the added advantage that the uranium content can be measured by its radioactivity without destroying any portion of what may be an extremely important find.

Another suggestion for an archaeological clock came from Professor H. C. Urey, one of the most fertile minds of American science, a man who for more than twenty years has kept the pot of earth sciences bubbling with a continuous output of new theories and ideas, and who has also been prominent in offering interpretations and ideas about the discoveries made on the moon. As far back as 1947 he showed that the ratio of the two oxygen isotopes, oxygen 16 and oxygen 18, in sea water varied according to temperature. Oxygen 16 is the common form of the oxygen atom; oxygen 18 is heavier, much less common, but a perfectly stable atom and not radioactive. Broadly speaking, the warmer the sea water the more there is of the heavy isotope, oxygen 18. Since this applies to sea water it also applies to water falling on the earth as rain or snow, as this atmospheric water is simply sea water that has been drawn up into the air by evaporation. Furthermore, quite small variations in the water temperature produce measurable differences in the ratio of the two isotopes of oxygen.

The idea of developing this oxygen-isotope ratio into a clock was furthered by more work by Urey and his associates in the early 1950s, but it was really taken up as a practical method by a Danish team led by W. Dansgaard of the University of Copenhagen. They pointed out that glacier ice should contain a continuous record of the temperature variations of the earth's climate, since each year's snowfall should contain the two oxygen isotopes in ratios that would show whether the average temperature of the sea water was increasing or decreasing. They put up this idea back in 1954, but it did not achieve any major results until help came from an entirely unexpected quarter – the U.S. Army Cold Region Research and Engineering Laboratory. The scientists of this unit in the late 1960s succeeded in drilling right through the icecap that covers Greenland until they reached the rock 1,400 metres below their base at Camp Century. From this drilling they recovered a complete and unbroken ice core, and the University of Copenhagen team have been able to study this frozen time-record. Later, the same U.S. Army Laboratory got a similar ice core from Byrd Station in the Antarctic, but for technical reasons, connected with the actual position of Byrd Station on the Antarctic ice, this second core has not proved so satisfactory a clock record as the Greenland core.

Since the waters of one half of the earth are warmer in that half's summer than in the following winter, the oxygen-isotope ratio will change from favouring oxygen 18 in the summer to a decrease of this heavier isotope in the winter. Thus the ice core can be counted off in years as the ratios of the two isotopes vary. But over the yearly variations the long-term climatic trends show too.

The one big snag of the ice-core clocks that has emerged so far is that the weight of ice on top gradually squashes and deforms the ice underneath. The whole depth of the Camp Century ice core probably represents the work of 100,000 years. Dansgaard and his team have shown that they can count back in individual years to about 6300 B.C. But for times before that the individual years tend to become squashed together too much, and only the long-term variations in climate are clear. The estimating of the correct number of years to count for a certain depth of ice as it becomes more and more squashed by the weight above then becomes an exercise in mathematics. The accuracy of dating further back than the time when individual years can be distinguished then becomes a function of the correctness of the mathematics.

Nevertheless, all the major variations in climate of the northern hemisphere can be clearly seen in the ice-record back to 100,000 years, and these can be correlated with the various increases and decreases of the ice sheets of the glacial period. These warmer and colder periods, with the ice sheets advancing or retreating across northern Europe and the northern United States, can be traced not only in the geological record and the botanical record on land, but can also be seen to have affected the cultures, activities and living areas of the prehistoric men who lived on these two continents.

Towards the top of the ice-records the yearly variations of climate can quite clearly be shown to correspond with known historical variations of climate; the 'little Ice Age' of Europe between 1600 and 1730 appears very plainly and also the long cold spell in the fifteenth century which is believed to have killed off the Norsemen's settlements in Greenland. Work on spelling out the record in the deep ice cores is still going on, and the value of the oxygen-isotope clock remains to be finally decided. But this clock system can also be applied to the deep-sea-bottom 'cores', which have been mentioned in connection with the radioactivity clocks. The minute sea creatures whose dead shells form the sediments on the sea bottom used the oxygen in the sea water to form the chemicals that made up their shells. The oxygen-isotope ratio of the time when any particular shell was formed is preserved within that shell, and a section or core through the sea-bottom mud therefore preserves a record of climatic variation at the surface far above. But there seems no method of finding a yearly variation in the sea bottom sediments, such as the ice provides.

An entirely different property of matter, magnetism, also provides the opportunity for constructing an archaeological clock and time-recording system. The system is called 'archaeomagnetism', and it differs from the systems above in more than just the property it uses. Archaeomagnetism should provide an extremely local clock but one of very high accuracy. It will never be a universal clock like radiocarbon and it will never be able to measure great sweeping events like ice ages – but it may be possible to pin down particular items and events of two thousand years ago to an accuracy of twenty years, which is far greater accuracy than any other clock can provide for that distance into the past.

It is known that the earth's magnetic field – the magnetic force which directs the compass needle to point north – varies continuously if measured at any one place, but the magnetic field is varying all the time in London in a different way from its variations in New York. At the moment the Magnetic North as seen from London is moving westwards about one degree in every five years. In the past it has wandered in the opposite direction, and this is known because magnetism was one of the first subjects that was studied in the earliest days of science in western Europe in the years before 1600. Much less well known to the layman is the fact that the magnetic field dips downwards quite steeply, so that a freely suspended compass needle will not only point northwards, but will also point downwards towards the ground at an angle of about sixty degrees. This inclination also varies and has been recorded as varying for many years. In theory, then, a compass needle in Roman London or in Mycenaean Greece or in modern Chicago will point at different angles both vertically and horizontally according to the place in which it is observed and according to the year in which it is observed.

Clay is the substance used for making bricks and pottery. Clay contains many particles of iron oxides of various sorts as well as particles of other chemicals formed with magnetic metals. Each of these particles can behave like a tiny compass needle if it is allowed to do so. In the normal state of cold clay, dug out of the ground, the metallic 'compass' particles are pointing all over the place and are held in their random positions by the structure of the clay. But if the clay is heated, as in the firing of pottery or brick, then the magnetic crystals become free to move. Each different iron oxide or other magnetic metal compound has what is called its 'Curie point' – this is the temperature above which it loses any magnetism it may have had before being heated. If the metal compound is then allowed to cool freely it will be affected by the earth's magnetic field at the place in which it is standing to cool, and it will become magnetized in the direction in which the earth's field is pointing at that place at that time. This magnetism will remain in the particle unless it is reheated above the Curie point again or unless it is subjected to some very strong remagnetization from a powerful magnet or electric field near by, which normally does not happen. The Curie points of most of the important metallic particles are in the range between 580 °C and 680 °C. Pottery is normally fired up to about 800 °C. So, in the making of pottery and brick, the clay used in making the pottery or the brick, and the kiln in which the firing is done, undergoes heating above the Curie point and steady cooling afterwards.

This means in turn that pots, bricks and kilns carry within themselves a record of the exact direction of the earth's magnetic field at the time and place at which they were last fired. Thus in theory if we can find the direction and the dip of the earth's field in a particular year at a particular place, and we can find pot, brick or kiln that came from exactly that place with the same magnetic directions, we can date the find.

In practice the kiln itself is usually the find on which the archaeologist concentrates because its reasonably certain that the kiln has stayed in that place and in that position ever since it was last used.

The theory of archaeomagnetism as a clock dates right back to the work in 1899, but it was only from 1933 onwards that serious and prolonged studies were started by the French scientist, Professor E. Thellier. It was not till thirty years after he started work that any wide circle of scientists and archaeologists became interested in Thellier's efforts. But now in Britain, Germany, Japan, Russia, Greece and Iceland, as well as in France itself, very serious efforts are being made to build up histories of the local variations in the earth's magnetic field, even as far back as 4000 B.C. in some areas. Archaeomagnetism promises to give the power of extremely accurate dating once the history, the basic chronometric plot, has been built up.

There are problems ahead, however. The work of building up the master-plot of the earth's magnetic field variation is tedious and lengthy; it has to start from some historically identifiable date and be extended backwards from there. No progress has been made in linking the variations in one place to the variations in another, though units the size of whole countries can be regarded as one place. There is not even a satisfactory theory as to how the variations occur, though it is believed that the whole of the earth's magnetic field is caused by electric currents in the fluid core of the planet and that local disturbances in the core cause local variations in the field above.

Nevertheless, it is already possible to date a kiln built in Roman London to A.D. 200 within fifty years accuracy and there is every hope that the range of accuracy will be halved in years ahead. Archaeomagnetism has many side-benefits in the possibility of obtaining accurate relative dates. It can, for instance, be applied to burnt-down buildings, which have of course been intensively heated during the conflagration or sack of a city. Thus archaeomagnetism could solve such problems of prehistory as whether the palaces of Mycenaean Greece were sacked and burned all within a very short time of each other in some terrible invasion or whether burnings took place well apart in time, and are therefore the remains of some slow decline.

Two further clock systems, both worked on by American teams, have recently been developed far enough to come into practical use for dating purposes – fission-track dating and obsidian dating.

Fission-track dating is a variation on the older theme of uranium-helium dating, and it is therefore one of the family of radioactive clock systems. But instead of counting, or attempting to measure, the total amount of helium present in a rock or mineral sample that has arisen from the release of alpha-particles by the decaying uranium, it counts the tracks made by the alpha-particles as they were shot out of the uranium. The helium may have disappeared by various routes during the ages, leading to inaccuracy in the dating. The tracks made by the alpha-particles will normally remain for thousands of years. These alpha-particle tracks are very short and quite invisible to the naked eye or even under the microscope. But the passage of an energetic alpha-particle changes the chemistry along its 'line of flight'. If a specimen rock is given a polished flat surface and then etching acid is poured on to it, the acid attacks any weak spots along the line of the alpha-particles' flight; this can then be seen under an ordinary microscope. The amount of uranium present in the sample is measured; its half-life is known and can be restated as the number of uranium atoms that can be expected to decay in a year; the number of tracks signifying an actual decay is counted; and the resultant calculation shows how many years have elapsed since the rock or mineral was formed.

Fission-track dating is virtually confined to the study of glassy materials or crystals. In practical terms this means that it is confined to those glasses formed in volcanic actions or man-made glass. Minerals formed by volcanic action are not, however, confined only to those areas where we now have volcanoes; the technique is eligible for far wider use than that, and it has been shown to provide results over a very wide time range from 'modern' to one thousand million years ago. Its most successful application so far has been to give an independent cross-check that the age of Bed 1 at Olduvai Gorge is around one and three-quarter million years old, confirming the age originally given by potassium-argon dating. Fission-track dating was only proposed in 1965, and it is, obviously, confined to specimens with an appreciable content of uranium, so it is at the moment having to prove whether it is widely valuable.

Obsidian is also a volcanic glass, but it was of enormous importance in virtually all Stone Age societies because flakes of obsidian had the sharpest cutting edge of any naturally obtainable material. Hence obsidian was exceptionally valuable to Stone Age men round the world – it made the finest knife blades in Japan, in the Old World and in the Americas, where the priests of the Toltecs used obsidian knives for cutting the hearts out of the victims on the top of the great pyramids which still stand today.

Obsidian comes in many colours, grey and green and black, but it has the peculiarity that the moment a surface of obsidian is exposed to the atmosphere – the moment man made a new knife blade, therefore – this glass starts absorbing water from its surroundings. The water slowly penetrates deeper into the glass and forms a 'hydration layer'. The depth of this hydration layer is proportional to the time the surface has been exposed to the atmosphere, and therefore there is a clock mechanism near the surface of one of the most common and most well-preserved of all Stone Age artefacts.

This possible 'clock' was only found to exist in the late 1950s, and it has been developed largely by Robert L. Smith and Irving Friedman of the Smithsonian Institute in Washington. The problems include the measurement of the hydration layer, which may be a few tenths of an inch but may be as little as millionths of an inch thick; the measurement must be accurate to less than ten millionths of an inch in order to get any sort of accuracy in the dating. There is also the difficulty that the rate of hydration, the rate at which water penetrates, varies with the temperature of the surroundings; it has been proved that obsidian exposed in the Arctic has a far thinner hydration layer than obsidian exposed on the hot coast of Ecuador. Knowledge of the local climatic variations over hundreds of years may therefore have to be included as corrections in the calculation of any dating. Nevertheless, many thousands of dates have been obtained from obsidian finds and have, on the whole, linked closely enough with known archaeological findings to allow some confidence to be placed in independent obsidian dates.

The importance of obsidian to Stone Age men provides both the importance and the archaeological difficulties in applying obsidian dating. This technique can only provide a local clock system because of the fact that the hydration layer varies according to temperature, yet obsidian is an extremely common archaeological find. It is particularly important up the western seaboard of the Americas, from Peru to California, and throughout the great Central American cultures. It was so valuable that it was an important trading item, and the trade routes of early American Indians can be traced largely through finds of obsidian. (The Stone Age in the Americas continued broadly until the arrival of the Europeans.) Obsidian was buried with great men as a sign of wealth and as a valuable thing to take to the next world. Herein lies the problem – it was so valuable that it was much used and used again. Hydration begins and the clock is started the moment the obsidian is flaked, while what the archaeologist wants is the date at which the great man was buried, or the date at which the obsidian was traded, and these dates are by no means necessarily the date at which the obsidian surface was exposed. The problem is an archaeological one and not likely to be solved by science.

All these clock systems have their own particular areas of relevance. Broadly, one can say that they all help out the radiocarbon method in some way or another – some by extending the range in years, some by providing a greater accuracy at some particular point, some by working successfully on the type of investigation where radiocarbon can never be of much help because there is little likelihood of any organic remains being found. But none of them have that universality and direct relation to the major problems of prehistory which have earned radiocarbon the title 'queen of the dating techniques'.

But now there has come one more 'clock' system which in many ways can rival radiocarbon dating both in universality and in relevance to prehistoric problems. This is thermoluminescence dating.

The original reason for people being attracted to the idea of using thermoluminescence as a system for archaeological dating was that this system is particularly applicable to pottery, and pottery is the true treasure found in archaeology. Jacquetta Hawkes wrote in praise of pottery:

> Of all prehistoric artefacts the most useful to the archaeologist for evidence of age and culture is pottery. This is for several reasons. To begin with, it has the essential quality of lastingness: any properly fired ware is indestructible in ordinary conditions. On the other hand it is easily breakable, so that on any site where people have lived there tends to be more of it than anything else. Again clay is very variable and slight differences in mixing, handling and firing it produce distinctive and readily identifiable wares; similarly it is a highly flexible medium

which allows an infinite range of forms, styles, modes of ornament. It is valuable again for quite a different reason; many objects, bronzes for instance, may be made by a few specialists, last for a long time, and be moved long distances about the map simply as articles of trade. In contrast prehistoric pottery is usually made by the women of each local community and is not easily transportable, nor is it all likely to have a long life before being broken; it therefore gives far more reliable evidence than bronzes for the movement or persistence of peoples . . . It is for all these reasons that pottery has come to have a very special place among the material documents of prehistory, and all archaeologists have to acquire some skill in reading it.[3]

If, therefore, the laboratory scientist could find a method of absolute dating for the humble, ubiquitous and telltale potsherd, it would be of the greatest significance for archaeology.

The fact of thermoluminescence had been observed in the seventeenth century, when Sir Robert Boyle noted in 1663 that certain substances, when heated up, give off more light than just the red-hot glow that might reasonably be expected. It is now known that this emission of light is shown by very many minerals when heated; the scientific explanation is that the minerals have, over the course of time, succeeded in trapping a number of extra electrons among the atoms making up their crystals, and these electrons are released from their traps when the crystals are excited by heating. The release of the electrons is a release of stored energy, and this energy is seen as light when the electrons escape. The energy got into the mineral crystals, or the electrons were trapped, because over the course of time the crystals have been exposed to energetic radiations from decaying radioactive atoms or from cosmic-ray debris particles coming down from the atmosphere. From this it follows that energy will be built up in the crystals over the course of years, and the thermoluminescence that appears when the crystal is heated is proportional to the over-all dose of radiation the crystal has received. The American, Farington Daniels, first showed in 1953 that this property could be used in a practical way, and the development of his idea has been the little metal badges worn by workers in the nuclear industries or by patients and nurses in radiotherapy departments of hospitals which measure the amount of radiation the wearer has been exposed to, and therefore act as a safety precaution.

But a Swiss group under Professor N. Grogler and an American team at the University of California, Professor George Kennedy and Dr L. Knopff, developed the idea towards an archaeological use. Their idea was that in the firing of pottery – even in primitive times temperatures in kilns up to 800 °C. were used – all the thermoluminescence in the grains of mineral matter in the clay would be released. Thus in the firing process the thermoluminescence clock would be set to zero; and all the thermoluminescence gained by the mineral crystals in the pot would be the result of radiation received after the firing. The amount of thermoluminescence that is shown by a piece of pottery heated now in a laboratory would be in direct relation to the length of time since the pot was fired. If the details could be worked out each pot would contain its own clock mechanism and the system should be applicable to every archaeological site in the world.

This idea was taken up in 1961 by a new group under Martin Aitken at the recently founded Research Laboratory for Archaeology and the History of Art at Oxford University. Theoretically the situation should have been simple. The pot whose date was wanted had had all its thermoluminescence removed when it was fired; assuming that it was broken and got buried within a short time, which Jacquetta Hawkes showed was one of the great advantages of pottery for the archaeologist, the amount of thermoluminescence it has gained up till the present day should be the measure of its age. The idea, then, was to measure the amount of thermoluminescence given off by the pottery when heated. Then by exposing the pot to measured doses of radiation from a controlled source of radioactivity in the laboratory till it produced the same amount of thermoluminescence, the total amount of radiation it had received in its lifetime, the 'accumulated radiation dose', was easily found. Then the quantity of radioactive elements in the pot itself and in the soil in which it had been buried could be measured.

The sources of radioactivity would normally be a few parts per million of radioactive uranium and thorium and a small percentage of radioactive potassium in the pottery itself and in the soil immediately adjacent to the pot. Because the half-life of all these is known the dose of radiation each year could easily be worked out once the amount of radioactive material had been measured. Then the mathematics was exceedingly simple – age of pot in years = accumulated radiation dose divided by dose per year.

But it took seven to eight years to get a workable method into operation. Their first attempts at dating were quite useless for the simple reason that grinding the pottery up so that it could be heated, measured and treated with radioactivity in practicable amounts and shapes put a 'false' thermoluminescence into it. They were in fact about to abandon the whole project when, from outside the group, came the suggestion that they should make their measurements in an atmosphere of nitrogen. Fortunately this proved exactly the way to cut out the false thermoluminescence. A fresh set of dates was measured on pottery whose real dates were known from other sources. This time the thermoluminescence technique got the pottery in the right order of date, even though the dates themselves were wrong, but they were wrong by a consistent factor. So they had at least got a satisfactory relative dating method.

It was theoretical work at Birmingham and Kyoto Universities which provided the necessary clues; it was realized that the thermoluminescence was carried by comparatively large particles of quartz and other minerals in the clay, while the radioactive sources, the uranium, thorium and potassium, were carried in the finer fabric of the clay, and since the range of radioactive effects is comparatively short in a dense material like clay, the evaluation of the radiation dose received by the particles that were responsible for the thermoluminescence was wrong. This problem has been overcome by developing two refinements in preparing the pottery for measurement. Now two sets of measurements can be made: one on very fine particles and one on comparatively large particles. It is hoped in the future that combinations of these measurements will eliminate many of the uncertainties, but the first real successes of thermoluminescence dating were achieved by using either the fine particle system or the coarse particle system alone.

Several other refinements and corrections have been introduced, to deal with problems such as the amount of water in the soil and in the pot during its burial period, and to give more accurate measurements of the radiation dose received by the pot from the soil around it. They have been enough to enable thermoluminescence to gain its first big success in providing dates for important samples of the pottery called 'Bandkeramik' from Neolithic sites in central Europe at Bylany, Heinheim and Stein. These sites had become a matter of major archaeological controversy by the late 1960s.

The traditional archaeological dating for these sites and the pottery they contained was 2500 to 3000 B.C., and these dates were based on the traditional method of a chain of relative dating which finally connected with objects that could be absolutely dated from the Aegean area and Egypt. The Czech archaeologist, Neustupny, had proposed a new 'long' chronology for central Europe in the Neolithic era which would put the sites back 1,500 years earlier – this controversy has been referred to in Chapter 4. The first radiocarbon dates for the sites certainly gave ages a thousand years, earlier than the traditional dates, and the radiocarbon dates corrected by the bristlecone pine chronology, put the sites well back a further six or seven hundred years to 4500 to 5000 B.C., fully supporting the long chronology. When the thermoluminescence technique was tried on the pottery for which these sites are famous there were the uncertainties which are to be expected from any new technique, but the results were consistent and unequivocal and they supported the long chronology and the corrected carbon dates to the full. In the serious atmosphere of an international symposium at the Royal Society in London, in 1970, Aitken said, '. . . The evidence, being from a completely different technique, should help to dissuade from their continued adherence to the short chronology those archaeologists who reject the radiocarbon dates (whether or not corrected).'

The thermoluminescence technique has been taken up by other laboratories; the well-known Museum Applied Science Center for Archeology of the University of Pennsylvania, for instance, has been working since 1968 on the techniques. It appears likely that eventually an accuracy of 5 per cent will be reached in dating pottery by this method using fairly complicated laboratory techniques. But cheaper and quicker thermoluminescence methods giving an accuracy of only about 15 per cent may be of more interest to most archaeologists except for special problems. It is also more than possible that thermoluminescence may be able to settle certain problems for radiocarbon dating over a stretch of years where the radiocarbon calibration curve is most difficult.

The establishment of any method of giving an absolute date to potsherds is bound to be of the greatest significance to archaeology, so the thermoluminescence dating technique is widely regarded as one of the most hopeful developments of the present time. But it promises even more, for the latest work, reported only in the last two years, shows that the method may be extended to provide absolute dates for at least some specimens of bones, teeth, and flint. This will give, for bones and teeth, a most valuable check on radiocarbon dates by an independent method, and if it can really cope with flint, then many Stone Age sites will be able to be dated to an accuracy never before visualized. Furthermore, those who question whether potassium-argon dating can truly 'link up' with radiocarbon dating have high hopes that thermoluminescence dating will be able to bridge the gap between the two. It should be possible to use thermoluminescence to date Early Stone Age fire-hearths and thus provide clock-times for one of the most obscure phases in the story of man's development.

1. J. and C. Hawkes, *Prehistoric Britain*, p. 162.
2. F. Johnson, 'Radiocarbon Dating and Archaeology in North America'.
3. J. and C. Hawkes, op. cit.

8

Revolution in a Revolution

In 1967 a twentieth-century American scientist, J. R. Harlan, made himself a rough sickle of wood with pieces of flint as the cutting edge. He based his sickle on the designs of similar implements that had been discovered by archaeologists at Neolithic sites in the Middle East. He took his sickle and went out into the large patch of wild wheat he had found on one side of a dry wadi in south-east Turkey. For exactly one hour he harvested the ears of the wild wheat in a way he imagined a Late Stone Age man might have done. At the end of that hour of work he had collected one kilogramme (about two and a half pounds) of perfectly satisfactory clean grain. Harlan's colleague, Daniel Zohary, witnessed the gathering of the ripe, wild wheat. He also saw Harlan draw the lessons of the exercise. Harlan imagined a family of Neolithic people in that area, people who had learned a certain amount about gathering wild grain. He knew that there was a three-week period during which the wild wheat became ripe and before all the ears had fallen naturally to the ground. His calculations showed that in that three-week period his imaginary Stone Age family could have gathered enough wild wheat to last them a whole year. 'Without even working very hard' (Harlan's words in his scientific report on the experiment) they could easily have gathered a ton of clean wheat.

The implications of this experiment were 'eye-opening' – again the description comes from a formal scientific paper. Clearly such a family could have lived permanently in that neighbourhood and lived off the wild wheat without ever going to the trouble of planting, weeding, watering and the other chores of agriculture.

Indeed, even if they hadn't been permanently settled there, how could they possibly have carried away a ton of wheat with them?

An anecdote? Yes – but an anecdote about a genuine, if not very complicated, scientific experiment. The results of Harlan's little exercise provided real scientific data which threw light on questions archaeologists had long been asking. It also exemplifies the impact which the biological sciences in general – those scientific disciplines which study plants and animals in their environments and in the laboratory – have had on archaeology and prehistory. It is a supreme example of the work of the bioscientists, which, together with the new results provided by the physicists' and chemists' 'clocks that run backward', has wrought a 'Revolution of the Neolithic Revolution'; those descriptive words that came from Dame Kathleen Kenyon, the excavator of Jericho.

Harlan and Zohary were primarily engaged, as plant-geneticists, in searching for the wild ancestors of the cultivated wheat and grain crops which have been the basis of all the European, Mediterranean and Middle Eastern civilizations since they were first domesticated. They had found, in south-west Turkey, that wild ancestors of our present wheats grew in profusion; as they wrote, 'over many thousand of hectares it would be possible to harvest wild wheat today from natural stands almost as dense as a cultivated wheat field'. This observation together with Harlan's amateur farming led them to ask, 'Domestication may not have taken place where the wild cereals were most abundant. Why should anyone cultivate a cereal where natural stands are almost as dense as a cultivated field? . . . Farming itself may have originated in areas adjacent to, rather than in, the regions of greatest abundance of wild cereals.'

The origin of farming is the central feature of that Neolithic Revolution which Gordon Childe had introduced to archaeological thought. In fairness to Childe it must be remembered that he visualized the Neolithic Revolution as being akin to the Industrial Revolution of the nineteenth century, rather than something sudden like the French Revolution; indeed he had the wisdom to see that the Neolithic Revolution must have taken thousands of years rather than the century of the Industrial Revolution. Nevertheless, the first concept of the Neolithic Revolution was a fairly simple one – that in some small area man had discovered the advantages of agriculture over the wandering, nomadic life of gathering food and hunting; that by conquest or example this idea had spread; that with agriculture came settlement in permanent villages which eventually grew into cities and large settled communities where writing and literacy were invented to cope with the administrative problems; that this was the origin of civilization; and that from this civilized centre in the Middle East all the stages of advance and progress diffused further and further outwards like ripples in a pool. The obvious importance of such an idea – its importance in the whole history of man and its importance as an intellectual concept in twentieth-century archaeology – inspired the archaeologists to turn their Middle Eastern excavation strategies away from the discovery of palaces and sculptures towards deeper diggings down to the Neolithic levels below. The great excavations at Mount Carmel, Jarmo and Jericho gave general confirmation to the thesis of the Neolithic Revolution.

But in addition to straightforward digging to discover the evidence for the earliest farming communities, in the shape of sickles or other implements such as querns for grinding cereals, and the traces of more or less permanent houses, there arose a number of other questions. Why had man discovered this new mode of living? What were the first crops? What were the first animals to be domesticated? Plainly botanists and zoologists at the least would be required to provide expert advice on such small traces of these things as the archaeologist could find. So in the 1950s and 1960s, just at the time when the physical chemists and other manipulators of archaeological clocks were making their influence felt, the bioscientists were also being called in to help.

Childe's own view of the cause of the Neolithic Revolution was that it was primarily a climatic change at the end of the Ice Age, giving a drier climate in the Middle East which threatened to turn large areas into desert or semi-desert. This change drove men up into the mountains where they learned to domesticate the wild wheats and barleys among the plants, and the sheep and goats among the animals. This view was broadly accepted until the intellectual ferment aroused by the advent of radiocarbon dating, which, as we have seen, pushed the origins of farming back before 7000 B.C. and attracted scientists from other fields into the discussion.

It is interesting to see that Childe's views were in tune with the comparatively simplistic 'evolutionary' theories of the biosciences of his time, and were also in accord with the Marxist outlook of his intellectual origins. The impact of the bioscientists, the laboratory dating experts, and the geologists and climatologists who have all been attracted by the problems of the Neolithic Revolution has been to give us a much less simple picture. The most recent work has made the Neolithic Revolution a much more complicated, more diffuse, more widespread affair – and incidentally it has made it much more interesting. Most modern views of man's discovery of agriculture see it in an ecological framework. Man was part of an eco-system, and he still remains so, though the eco-system has changed; there are changes in plants and animals just as much as there are changes in man. There is no such simple thing as 'agriculture'; there have been changes in the various different types of eco-system of which man is part, and his methods of cultivation and domestication, his food habits and requirements have been different in each system. This ecological view is just as much in tune with the general scientific trends and fashions of our times as Childe's Marxist and evolutionary view was typical of the climate of opinion forty years ago. One may therefore wonder whether views about the Neolithic Revolution will change again in the future as the general trend of scientific thought takes some new emphasis.

It is also in accordance with the experimental and scientific emphasis which is now pervading archaeology that at experimental farms at Churchfield in England and at Lejre in Holland archaeologists are using wooden ards and cattle-drawn tools to establish the efficiency of earlier agricultural techniques. They even grow primitive wheats in their gardens to establish their output and fertility. Only after this work can calculations about primitive economies be realistic.

The traditional view of the Neolithic Revolution can be criticized as being an example of 'The Whig Interpretation of History', a pejorative phrase coined by Professor Herbert Butterfield, a Cambridge historian who specialized in English history of the seventeenth and eighteenth centuries. It means here the adoption of the idea, unconsciously, that 'progress' and 'development' were those steps which we can see have led to that pinnacle of achievement, ourselves; seeing the development of agriculture as leading to those 'affluent' arts of pottery and weaving and hence to 'civilization'.

It was the anthropologists who first upset this view. They pointed out that hunter-gatherer societies were not necessarily impoverished; by no means without leisure expressions of the human spirit in art; not desperately spending every moment of their time in grubbing a bare existence from their environment. The clearest examples were the American Indian societies of the north-west coast, mainly in what is now Canada. Technically these were Stone-Age men when first 'discovered' by Europeans, but their lives were built around the super-abundant crops of salmon and other fish of that area. These people had permanent and substantial wooden houses, much decorative art (their famous totem poles are the obvious specimens), and so much affluence that a feature of their community life was the deliberate and ceremonial destruction of collections of material which, to them, represented wealth. In Tanzania the present-day women of the Hadza tribe have been shown to need only two hours a day, throughout the year, for all the activities necessary for subsistence. The Bushmen of southern Africa do not use all the resources available to them; they deliberately refrain from collecting certain types of edible berries for cultural reasons only. It was the archaeologists themselves who had discovered the cave-paintings of southern France and northern Spain, paintings which are a superb art-form, but done by men who were 'primitive' hunter-gatherers.

The anthropologists likewise pointed out that there were other forms of agriculture than the growing of cereal crops which marked the Neolithic Revolution. There is the wet-paddy rice-growing system of eastern Asia, there is the yam-and-tuber agriculture of central Africa and south America, which is more like horticulture, the gardening of plots near the home, or is based on growing crops in artificial mounds. And there is the maize agriculture of Central America. Each of these different forms of agriculture or crop-growing seems likely to have sprung independently from different ecological systems which had little in common except that man was one of the inhabitants.

Professional ecologists asked rather different questions about the origins of agriculture – believing that good science is based on asking the right questions. They asked why, out of all the types of plants available, did men start agriculture with wheat and barley? Why sheep and goats rather than gazelles for the first herds? Why did agriculture originate only in certain restricted parts of the world – south-west Asia, China and Central America? It is quite clear that man exploited many different plants when he was in the collecting phase; this is shown most dramatically by the stomach contents of Tollund Man – recovered whole from a Danish bog two thousand years after his death – which included not only cultivated cereal grains but a wide variety of seeds from wild plants, some in such large proportions that they must have been collected deliberately. Even in our automated, machine-dominated western cities we still go out to collect blackberries or hazel nuts. So despite the fact that humans have collected the fruit or seeds of many plants throughout the ages, most of the plants have remained wild, and only a tiny number have become cultivated. To the ecologist it seems plain that the particular plants that were eventually cultivated must have 'needed' man as much as man needed the plants, for in the ecologists' eyes man is only as much part of the eco-system as the plants are.

The explanation put forward at the most recent international symposium on this subject is that the ancestors of the cultivated plants were 'weedy', in the broadest sense of that word, and not in the gardener's sense of 'unwanted'. In this view, most plants are regular inhabitants of stable eco-systems, the plants of the forest or of the grasslands or of some other system. 'Weedy' plants, however, are those that have evolved so as to seize their opportunities in broken ground, in disturbed areas or in open places where there is unstable, bare, soil with not much competition from other plants, which have evolved to grow in dense, stable plant-communities. Yet even among 'weedy' plants, only some have been turned into crops, and the suggestion is that the crop plants were precisely those that benefited most from man's activities, from the open disturbed soil around the temporary settlements of men, and from the richness of the soil fertilized by men's detritus; 'such plants sought man out as much as he sought them out – because of their specific manurial requirements.'

How did the ancestors of the cultivated cereals survive in pre-agricultural times? To avoid competition with other plants it is evident that they could only grow and survive in poor, thin, soil on rocks, amongst stones or in sands and gravels. Furthermore it seems necessary for them to have been confined to regions with a well-marked wet and dry season, since if the rainfall was continuous such areas would quickly become covered by lush rain forest or dense grasses. These plants, ancestors of our cultivated plants, were opportunists; they needed to germinate and grow quickly when the rains came in the spring and when the ground warmed up, but equally they needed to complete a full life cycle and mature their seeds before the ground dried out in summer. Thereafter the seeds lay dormant in the soil, germinating perhaps a little in the autumn rains and growing again in the spring to set a new crop of seeds in the early summer.
. . . The seeds needed to survive the long hot dry season in a well-baked thin soil and there must consequently have been strong selection pressure for large seeds with large food reserves to resist the drying out and grow quickly when the rains came again. In these soils and under these conditions nothing with small seeds would survive well, but nor would large perennial plants either, so these ecological weeds, the ances-

tors of our cultivated plants, were able to grow and survive under these special conditions.[1]

This sort of plant – weedy but with large food reserves in their seeds – found ideal conditions in men's rubbish heaps and therefore colonized them. But these were exactly the plants whose seeds man would gather in preference to others with less food in them. To quote Hawkes again, 'To primitive man it must have seemed little short of miraculous to find that plants needed for food sprang up by his very huts and paths. Perhaps it is not too far-fetched to suggest that this situation may have been the basis for so many folk-legends which attributed the beginnings of agriculture and the introduction of useful plants to gods or supernatural beings.'

If the evidence of the anthropologists and the views of the ecologists are put together with the recent discoveries by the climatologists that south-west Asia actually became more humid in climate as the Ice Ages drew to a close, a new suggestion for the origins of agriculture, at least in that area, can be easily supported. The suggestion by Lewis Binford is that there was a population expansion among the settled hunter-gatherers of the best areas of south-west Asia; these were broadly the Natufians of the caves of Mount Carmel and the bottom layer of Jericho (referred to in Chapter 1). The expansion of population drove some groups out to the marginal areas, where they could no longer find stands of wild wheat, such as Harlan harvested in this present decade, and they had to develop the technique of storing seed and sowing it deliberately to 'keep up the standard of luxury to which they were accustomed'.

This conclusion has been reached by applying what might be termed a 'scientific' rather than an 'archaeological' approach. But it has considerable support from the archaeological record and from the opinions of certain archaeologists who have concentrated on early sites in the Middle East such as Professor Jean Perrot, of the French Archaeological Mission in Israel. Perrot has excavated the site of Mallaha near the Israel–Lebanon border and found a village type of site, dated to around 9000 B.C. with a definitely Natufian culture, but with no signs of food-production or agriculture. The people of Mallaha would nowadays be described as in the 'intensive food-collecting stage' which is held to be one of the four steps from hunter-gatherer ways of life to the full agricultural life.

The archaeological record has also provided the first clues as to which plants were the ancestors of our modern cereals and other crop-plants. The number of ways in which plant remains are preserved in the archaeological record is quite surprising. Whole grains and many parts of the grain or husk have been found embedded in pottery. There are also many cases in which the wet pot has received the imprint of a grain of wheat and retained it during firing so that the record remains clear to this day. Since pottery was usually made by the women in the living area these finds are not unexpected. The mud-bricks of the Middle East, so often strengthened by straw, similarly often contain grains or the impressions of grains. But even more preservative was the ancient practice of heating or parching cereal grains. This was necessary to get the grain out of the tough husks, which, in the primitive wheats, could not be broken open by the mere mechanical impact of threshing, or grinding. The heating process was not always well controlled: 'Thus it often happened that the grain was overdone, scorched or wholly carbonized. Then it was thrown on the midden – and was preserved for ever.' The mummy grain of Egypt, found in passages beneath the pyramids or in storage pits at ancient sites now in the desert, is the finest record of all. Here the grain has been preserved by sheer dryness and it has been preserved in such a way that it represents the most accurate record we can possibly get of the original food plant. (Incidentally the story that mummy grain will sprout again is a complete myth – the very dryness that has preserved its form has also completely killed it.)

But to identify these archaeological finds is the work of the botanist, or at least of a scientist trained in botany who may nowadays be a specialist palaeoethnobotanist. One man more than any other has unravelled the history of the first crops – the

Dane, Hans Helbaek of Copenhagen University. He has done it by sheer hard work, carefully comparing the grains, spikelets, husks and stems of plants preserved and found by the archaeologist, and also comparing them in pattern and measurement with the corresponding parts of plants that grow today.

This work has established clearly that the first three crop plants domesticated by man in the Middle East were two types of wheat, einkorn wheat, which is scientifically called *Triticum monococcom*; emmer wheat, *Triticum dicoccum*, and one type of barley, the two-row barley, *Hordeum vulgare*.

But to find the wild ancestors of these crop plants and to establish how they were developed into the modern plants the plant-geneticists and plant-breeders had to be called in. They have proved that the cultivated version of einkorn wheat, *Triticum monococcum*, must be descended from the wild einkorn, *Triticum boeoticum*. Both plants have the same number of chromosomes (fourteen, regarded as two sets of seven chromosomes). They can be crossbred and the resultant plants are fully fertile. The wild einkorn evolved in the Fertile Crescent, though it did not populate the whole of that area, but was concentrated in southern Turkey, northern Iraq and northern Syria. Perhaps significantly, it has spread to other areas as a weed, in the modern sense of that word, and seems possibly to have followed, as a weed, the spread of cultivated einkorn.

Similarly, cultivated emmer wheat must have originated in wild emmer wheat, *Triticum dicoccoides*. Both plants still exist and both have twenty-eight chromosomes, which are regarded as two sets of fourteen similar chromosomes; the plants likewise produce fully fertile offspring when crossbred. It was plainly from this line that the 'durum' wheats developed, *Triticum durum*. These are 'naked' wheats, more highly evolved and much easier for primitive men to husk and grind into food. The wild ancestor here has a much more limited range and began its life in the area of Palestine, southern Syria, the Jordan Valley and across into Trans-Jordan.

These wheats are not our modern western bread wheats,

however. Bread wheat, *Triticum aestivum*, is a different plant. It is never found in a wild form, and the work of Helbaek, using the archaeological record, shows that it only appears one or two thousand years after the first arrival of emmer and einkorn wheats. *Triticum aestivum* has been proved by the plant geneticists to be a result of hybridizing the emmer-durum lines of wheat with the 'goat-face grass', *Aegilops squarrosa.* This plant is of the same family as the wheats in broad terms, and it has fourteen chromosomes, in two sets of seven. Modern bread wheat has forty-two chromosomes, which can be regarded as the twenty-eight chromosomes of emmer wheat hybridized with the fourteen chromosomes of goat-face grass.

But the discovery of the origin of bread wheat in a fusion of emmer wheat and goat-face grass tells a great deal more than is obvious at first sight. The goat-face grass in its original wild state flourished not only on the easternmost portion of the Fertile Crescent but spread well up into the mountains that separate India from Russia; it is a much hardier plant than the emmer wheat which evolved to suit the warm Mediterranean climate of its homeland. The homelands of emmer wheat and goat-face grass do not intersect, and hybrids between wild emmer wheat and goat-face grass are weak and dwarf plants. Hybrids between cultivated emmer wheat and goat-face grass, however, are strong bread wheats. From this it follows that the forty-two-chromosome bread wheats can only have been formed after man had started agriculture and carried the knowledge of cultivated emmer wheat into the homeland of goat-face grass. The hybridization between the grass and the wheat must have been accidental, but is not surprising for the grass is a persistent and aggressive weed in modern Middle Eastern wheatfields. But once the hybridization had occurred and had been taken up by early farmers and encouraged, then the hardy qualities of the grass enabled man to spread the growth of wheat into areas and climates in which emmer alone would never have survived.

Barley has a much simpler ancestry. It comes from one ancestor only, the wild barley, *Hordeum spontaneum*, a two-row barley which had a wide original habitat from the eastern Mediterranean, including all the Fertile Crescent to as far east as Afghanistan. The archaeological record shows clearly that early farmers had a two-row cultivated barley, called *Hordeum vulgare*. Both the wild and cultivated barley can be crossbred with each other, and both have fourteen chromosomes, two sets of seven. Archaeology shows that a modern form of barley, the much more productive six-row plant, came into use in the Middle East later than the two-row plants, and this six-row plant is the ancestor of our modern crop. Plant genetics has shown, however, that our modern *Hordeum vulgare* of the six-row type is the same species as the ancient two-row *Hordeum vulgare*, and the six-row feature must have come about by a natural mutation which was seized upon and encouraged by early farmers. A wild six-row barley found as a weed in Middle Eastern countries, *Hordeum agriocrithon*, which was once thought to be the ancestor of six-row cultivated barley, has been shown to be not truly wild, not capable of survival on its own, and therefore a derivative of the cultivated six-row plant.

There is, however, an important corollary to this matter of the development of plants into forms suitable for man's agriculture: plants suitable for agriculture cease to be able to exist in the wild state. There is therefore an important difference between a cultivated plant and a domesticated plant; a cultivated plant can continue to exist if man stops cultivating it, whereas a domesticated plant would simply die out if man ceased to provide the ecological conditions in which it can grow. The domesticated plant has become as dependent on man as man is dependent on the plant. This is how Helbaek expresses it:

> Many results of mutation are not viable under the ecological conditions obtaining where the change takes place, and thus the mutant, even if not sterile, will die out from environmental pressure. However, if man improves growth conditions for a species, more individuals may be accommodated in the same area unit, and this enables some mutants, that in the untended state would have perished from competition, to survive in consequence of 'relaxed selection pressure'. If attractive, such mutants may be picked out for special attention, and at that moment their status is changed: they cannot return to an independent

existence in nature; they are domesticated, that is, tied to men. . . . In the independent state the spike of wild barley disintegrates at maturity and the units are dispersed by wind or animal transport. The units, or spike sections, are admirably built for wind transport, of an aerodynamic design reminiscent of a modern aircraft and with appendages for anchoring and self-planting. Only the median (middle) of the three florets bears a grain, the two empty lateral organs act as stabilizers. The moment a mutation happens making the two lateral florets loaded with developed grains, the unit becomes a clumsy structure that will fly no more than a bulldozer; it drops limply to the ground and, nota bene, all the units will drop in the same place, creating a quite impossible situation of growth competition. If man takes a hand, this otherwise hopeless impasse can be overcome, and then only in a useful way if another mutation happens, keeping the spike axis articulate at maturity and not disintegrating. Consequently, barley having suffered these two physiological changes would be absolutely helpless in nature; but would in the hands of man, yield a large surplus over and above the grain necessary to reproduce the field with the same or even increased density of individuals. Most conspicuous is, of course, the case of naked-grained six-row barley.

This is the fundamental meaning of domestication. So it may be concluded that a cultivated plant need not necessarily be domesticated – cannot by any means be so from the outset – while on the other hand a domesticated plant can exist only as a cultivated plant. Cultivation is a matter of directing ecology, whereas domestication depends on some physiological inefficiency in a plant, advantageous to man.[2]

But man needs a balanced diet, a combination of protein, starch and fat. In the strict scientific sense, he cannot live by bread alone. Either instinct or experience had taught this fact to primitive man, and, although we know it well enough, it is all too easy to forget that our ancestors knew it as well. Forgetting that they knew it can lead to a misreading of the archaeological record, and this may be brought home to us in unexpected ways. Kent Flannery describes how he was sitting with his Rice University colleague and leader Frank Hole by their excavation in the Khorramabad valley in Persia in 1963 sharing a paper bag of local pistachio nuts as the final course of their evening meal. Like many local products these pistachios were delicious but fraught with some danger in the shape of many small, living caterpillars which had riddled the nuts. So Flannery examined each pistachio carefully as he took it out of the bag and opened it and he had to discard almost half of those he got from the bag. Then he noticed that Frank Hole was able to eat every single one of the pistachios he picked from the bag. Flannery commented on his good luck and Hole replied, 'I'm just not looking.' Like a good scientist Flannery did not let the remark pass. He later discovered that one kilo of dried caterpillars contained more than 3,700 calories, 550 grammes of protein as well as calcium, thiamine, riboflavin and iron. Indeed the protein content of caterpillars is considerably higher than the protein in pistachios, and there seems little doubt that caterpillars-and-pistachios give a better diet than pistachios alone. Flannery ends his cautionary tale with the words, 'Hole's wise decision to diversify his subsistence base brought him out of the field season a good thirty pounds heavier than me.'

But Hole and Flannery are distinguished for more than just good humour in the field. Their exploration of Deh Luran in Persian Khuzistan was the first to show in any detail a real picture of what actually happened during the process of developing agriculture. They succeeded in this largely because they persuaded Hans Helbaek to join their dig.

Deh Luran is a small plain in the foothills of the Zagros Mountains of western Persia. It is near the south-eastern tip of the Fertile Crescent, and is separated from Iraq, the land of the Tigris and Euphrates, the land of the first civilizations, only by a low ridge of mountain on its western side. It is treeless but has some vegetation in the shape of bushes and plants. It is not desert, it gets some rain, but it is pretty bleak. The temperature rises to well above 50 °C., (100 °F.) in the blazing sun of summer, but ice forms on the water buckets during December nights. There is a swamp in one part of the plain, obviously the remains of a former lake; there are many dried-up watercourses, and some areas are so boulder-strewn that driving a vehicle is impossible. Yet the whole plain otherwise is littered with the mounds that tell of former inhabited villages, some of which are clearly as old as Neolithic in date.

The excavation began in the autumn of 1963 and work went on until the main results were published in 1969. The team dug two mounds, one on each side of the swampy depression in the plain. First was Tepe Ali Kosh where the first settlement (the bottom layers), called the Bus Mordeh Phase, dated back to 7500 B.C. This was followed by a clearly distinguishable second phase, named Ali Kosh. The top, or latest, phase ran on up to 5600 B.C. where it petered out, as the culture they called Mohammed Jaffar appeared to end in slow decline and degeneracy. The second mound, Tepe Sabz, picked up in time almost exactly where the first left off, but plainly belonged to an entirely new culture, a different set of invading people.

The technique used for finding the botanical archaeological record involved truly extraordinary labour and thoroughness. Every spadeful of 'interesting' material was subjected to the 'flotation' method. First the material was dried in the shade and then it was put into water and allowed to settle; this meant that the lighter plant material floated to the top while all the earth and mineral pieces sank to the bottom. The water was then poured off and the plant material was caught in a sieve. Helbaek records his thanks to Frank Hole, the director of the excavation, simply for his generosity with water: 'all the water we sloshed away on the floating, every pint of which was brought in by truck and dearly paid for.' And of others he writes: 'The back-breaking job of the actual floating was taken over cheerfully by the hard-pressed Kent and Nancy Flannery in their so-called "spare time".' Helbaek's own task, however, was to examine under the microscope literally millions of small bits and pieces, from which he picked out more than 45,000 grains, seeds and spike fragments which he could identify to some extent, at least stating which family of plants they came from.

The archaeological record of seeds and plant remains shows clearly that the first inhabitants of Ali Kosh, the Bus Mordeh people, came into the valley from outside, and they brought with them domesticated emmer wheat and six-row barley. They also had flocks of domesticated sheep and goats, according to the finds of bones, though mixed with these were the bones of large numbers of wild animals they had hunted and killed, and the traces of fish and freshwater mussels. But the really surprising thing that emerged from the laborious floating technique and the patient counting of everything that was recovered was that these people relied on their crops for only a small proportion of their plant food. Two thirds of the total weight of plant food they ate must have come from wild legumes (plants of the pea and bean families) and from the seeds of wild grasses. In fact if we calculate by the number of seeds they left behind in their rubbish dumps for Hole and Flannery and Helbaek to find, 95 per cent of the total were the seeds of wild plants while only 5 per cent were from cultivated grains. However, the individual grains of emmer and barley are so much larger than the seeds of wild plants that they make up to 30 per cent when the weight of food is calculated.

Of this extraordinary discovery Helbaek writes:

> Treatises theorizing upon the plant food of Palaeolithic man consistently emphasize the large-grained wild barley and wheat as his primary or exclusive goal. The Bus Mordeh Phase material makes a considerable dent in this one-sided concept . . . Of course man would go for easily collected plant food, and wheat and barley would be obvious targets where they occurred, but here, quite unexpectedly, we find an enormous labour reflected in the collection of small plants with tiny seeds, exceedingly difficult to unhusk. Evidently the early collector-farmers recognized the importance of a balance between starch (grasses) and plant protein (leguminous seeds) in a healthy diet and this lesson was hardly learned on the salty arid plain immediately upon their arrival in that novel ecological world. It certainly reflects very ancient experiences in regard to plant food.[3]

The picture therefore is not the expected one; it is of a tribe who were part-hunters, part-herdsmen, part-farmers and part-food-gatherers. In the next phase, the Ali Kosh Phase, the amount of cultivated grain found in the remains goes up to over half, suggesting that this people depended much more heavily on their agriculture. Because the remains of the Bus Mordeh phase include a few seeds of wild plants that favour waterlogged ground, and these seeds had not appeared in the very earliest layers, Helbaek suggests that by this time fairly regular 'fields' or plantations along the edge of the lake were being used.

Another surprise comes in the third stage, the Mohammed Jaffar Phase. Here the amount of food from cultivated crops decreases to almost nothing. There is an increased quantity of meat and a vast collection of wild plants from the steppes. Field weeds such as grasses virtually disappear from the record, but seeds of plantain and mallow, which had undoubtedly been available on the steppes throughout the centuries of occupation, suddenly appear in large numbers. These latter plants favour disturbed, but not actually tilled, land and occur in and around tracks and goat paths. The implication is that the lake of the Deh Luran plain had shrunk and receded from the village until agriculture was not worth while. Then the village was abandoned.

But in this last phase there are several significant small features. There is the first appearance, although only two grains, of the modern domesticated barley plant, *Hordeum vulgare*, the six-rowed easy-threshing plant. All former traces of barley had been the six-rowed version of the original wild *Hordeum spontaneum*, the cultivated, but not domesticated, barley. All three early phases had shown a few samples of flax-seed, one of the most valuable crop-plants in later times for its oil and linen. Now flax cannot grow in the natural conditions of the Deh Luran plain, so these specimens must have been brought in from the mountains of the north as trade imports, but each time experiments with growing the crop must have failed. Goat-face grass, one of the ancestors of bread wheat, also appeared among the seeds. But interestingly there are no remains of pistachio nuts, although they grow in the mountains above Deh Luran; the only explanation must be that there were fierce mountain tribes at war with the comparatively settled farmers of the plain.

The desertion of Ali Kosh occurred at about the same time as the settlement of Tepe Sabz was started. But Tepe Sabz was started by a new foreign people who arrived with a much wider selection of crops based on irrigation techniques. They also had pottery. They did not settle near the lake; they dammed the stream beds where the water comes off the mountains in the spring, and later they carried this trapped water to other places by canal. They were people who had learned their agriculture by riversides or stream banks.

They had flax-seeds of a very large size, much bigger than those grown down in the plains of Iraq before irrigation was invented, seeds as big as any found up till 3000 B.C. and then only found in areas of irrigation-agriculture. Since flax will not grow naturally in Deh Luran, these people must have practised irrigation. They also had bread wheat, *Triticum aestivum*, cultivated emmer and einkorn wheat and the modern form of six-row barley. This latter again points to irrigation. It appeared briefly in the earlier stages but never managed to survive. At Tepe Sabz it is a regular crop. The earlier appearance may represent one natural mutation or a deliberate attempt to introduce a mutation that had been developed elsewhere. But in the plains below Deh Luran the modern six-row plant makes its appearance wherever irrigation seems to have been started. The evidence, then, is clear; the new people who came to Deh Luran about 5500 B.C. came up from the plains below, bringing with them their irrigation agriculture, their better and larger selection of crops. They brought, too, the first domestic cattle that Deh Luran had seen. They were farmers in the proper sense of the word, and the developments they brought with them had occurred down in the plains probably nearly a thousand years before they were brought up to Deh Luran.

But the new people also brought with them the seeds of their own destruction. Their irrigation methods eventually destroyed the fertility of the land by depositing natural mineral salts. There have been later dwellers in Deh Luran, at the time of the Persian empires and in early Islamic days, but it is probably true to say that Deh Luran is less inhabited now than it was in 7000 B.C. Yet this small, bleak plain provided us with our first view of what actually happened during the two or three thousand years of the Neolithic Revolution.

This outline has given a picture of steady progress in the depth and range of our knowledge of prehistoric times and in particular of the most important development in man's history. This progress has been achieved by bringing other scientific disciplines to the aid of archaeology. But it would be wrong to paint a picture of unalloyed success in the partnership between the laboratory scientist and the archaeologist. One notable area in which there has been failure is in the attempt by physiologists and zoologists to find out exactly what identifiable changes took place in the process of domestication of animals.

It is not always possible to tell whether a piece of bone unearthed by an archaeologist is definitely that of a wild animal or definitely that of a domesticated species closely related to our modern farm animals. There is a large area, in archaeological terms large, in which we cannot tell whether the men in a particular settlement have herded animals or hunted them. Basically this is because there is usually no essential physiological difference between the wild sheep, pig, goat or cow, and the domesticated version, at least in those organs which are likely to be preserved for thousands of years. We can tell from which wild breeds our domesticated animals are descended and, for the Old World at least, this points again to the Fertile Crescent or the mountains immediately north of it for the earliest domestication of the sheep and goat. Cattle and pig present a more difficult problem, and so too does the dog. For instance, the skulls of what might be the first domesticated dogs were found at Jericho, but when closely examined by Juliet Clutton Brock it seemed just as likely that the remains came from a small Arabian wolf. Size is the only criterion usually left to the zoologist, and the most recent studies have shown that large individual domestic cattle of early times were sometimes bigger than individual small wild cattle. This is complicated by the possibility that female wild cattle may have been much smaller than males and overlapped in their size-range by large domesticated bulls. The only convincing evidence often comes from an analysis of the age of all the animals killed at one site in one period. If a high proportion are in early maturity there is an implication that the creatures were being deliberately herded and killed for food as soon as they reached an economic size.

The problems of deciphering the record of the Neolithic Revolution are not yet solved. The situation was well summed up by Peter Ucko and Professor G. Dimbleby in the Preface to the collected papers of the International Symposium on 'The Domestication and Exploitation of Plants and Animals':

The domestication of plants and animals was one of the greatest steps forward taken by mankind, and although it was first achieved so long ago we still need to know what led to it and how, and even when, it took place. Only when we have this understanding will we be able to appreciate fully the social and economic consequences of this step. Even more important, an understanding of this achievement is basic to any insight into modern man's relationship to his habitat. In the last decade or two a change in methods of investigating these events has taken place, due to the mutual realization by archaeologists and natural scientists that each held part of the key and neither alone had the whole.[4]

Scores of excavations throughout the 1960s, and still going on, are attempting to shed more light on the beginnings of agriculture, pottery and village life. They spread from the valleys of the Danube and the Nile right through Turkey and the Middle Eastern countries, Persia, and as far as Afghanistan to the east; and there is much work in the southern areas of the U.S.S.R. too. It may be unwise to concentrate so much effort on what is only one of at least three areas in which agriculture was 'invented' independently, but most scientists and archaeologists belong to those societies whose cultures stem eventually from the Fertile Crescent. In fact there were other Neolithic Revolutions, though they are not usually called so, and the most clearly exposed is the one that took place in Central America.

1. J. G. Hawkes, 'The Ecological Background of Plant Domestication'.
2. H. Helbaek, 'Plant Collecting, Dry Farming and Agriculture', p. 404.
3. H. Helbaek, 'Prehistory and Human Ecology of the Deh Luran Plain'.
4. P. Ucko and G. Dimbleby, *The Domestication and Exploitation of Plants and Animals*, preface.

9

Man and Maize

It was the search for the origin of maize that led to what is arguably the most important single archaeological expedition in the history of the Americas. The expedition was to the valley of Tehuacan in central Mexico and it changed our views about the origins of agriculture, about the prehistory of the great civilizations of Central America, about the Neolithic Revolution itself, and even about the origins of man in the Americas.

Maize is probably the most important single food crop in the world nowadays; it is certainly the most important crop in the Americas. Maize was the foundation on which rested the great cultures of the Mayas, the Aztecs and the Incas. Yet the plant was quite unknown in the Old World before Columbus 'discovered' the New World in 1492. And, whereas we can still see the wild ancestors of our Old World food crops of wheat, barley, or oats, there is no known wild ancestor of maize.

The mystery of the origin of maize added to the general mystery about the origin of man in America. Setting aside the suggestions that the American Indians originated from the Ten Lost Tribes of Israel, or from the Egyptians, or from the inhabitants of the sunken continents of Atlantis or Mu, it was generally agreed scientifically that the aboriginal inhabitants of all the Americas came from Asia by way of the Bering Straits, where the geologists provided evidence that there had formerly been a land-bridge between Siberia and Alaska. The earliest evidence of human occupation of the Americas seemed to date to about 10,000 B.C. or rather earlier, but at sites of this date such as Folsom, Clovis and Sandia, the stone arrow-heads were contemporary with the bones of mammoths and other creatures which had died out in the Old World thousands of years earlier. Radiocarbon dating settled this problem and made it clear that the animals had simply survived longer in the New World. In general, the spread of datings from the far north, at about 15,000 B.C., to the extreme south, where stone artefacts remarkably similar to the Folsom types could be put at about 10,000 B.C., told a satisfactory story of steady movement by the original immigrants and their successors from north to south at a time when the Ice Age was ending.

Nevertheless there were disturbing features in this simple picture. First, why had they gone north to south when the ice was moving north? Secondly, the geologists kept wanting to put the date of the land-bridge between Asia and Alaska earlier. Thirdly, there were disturbing examples of cultural 'freaks' on the west coast of America which could have spelt contact with more advanced East Asian civilizations – items such as wheeled toys, although the Aztecs and Incas never used the wheel, and pottery styles on the Peruvian coast which looked remarkably like early Japanese. There were finds of stone tools more primitive than those of the mammoth-hunters, implying an earlier set of cultures in the Americas. Finally there was a skull found in Ecuador, nicknamed 'Fred', and radiocarbon dated to 28,000 years old.

If there was a question about when the immigrants came, there were even more questions about what they brought with them. In particular did they bring agriculture and maize with them? It is now clear that there was no one simple immigration. There were many different racial stocks and languages among the Amerindians when the Europeans 'found' them. (Anthropologists can claim as many as 8 racial groups, 125 different languages and 600 dialects.) There is no trace of maize in Asia, nor was there any sign of rice, the staple crop of Asia, in America when the Europeans arrived.

The argument, as so often in archaeology, boils down to a battle between the diffusionists and those who hold that man is perfectly capable of making the same invention, perhaps over and over again, quite independently at different places and different times.

This argument continues vigorously today, concerning not only the development of agriculture in America, but the invention of the wheel, the source of Chinese civilization and many other subjects. For the diffusionists, there are the practical experiments of Thor Heyerdahl who has sailed across the Atlantic on a papyrus boat, and across the Pacific on a balsa raft; and above all there is the argument of the diffusion of ideas which implies that, though the Chinese and the American Indians may have domesticated plants and started agriculture themselves with their own native plants, the idea of agriculture, and the benefits to be gained from having a steady food supply, could have reached them by word of mouth, by the most casual or chance contacts with individual traders or wanderers from other parts of the world.

But the strongest anti-diffusionist answers came from the hunt for the origin of maize. Perhaps more important for the purposes of this book, that hunt brought forth the first major interdisciplinary attempt in the archaeology of Central America, the first time that botanists, zoologists, geologists, radiocarbon-daters, ecologists, climatologists and others had been deliberately, and with forethought, brought into an archaeological exploration. The moving spirit in the hunt was Richard S. MacNeish, who started his work for the Department of Anthropology at the University of Chicago, but is now with the Robert S. Peabody Foundation for Archeology at Andover, Massachusetts.

He started in 1945, when it was generally believed that man in America dated from 10,000 B.C., and, consequently, that the cultivation of maize had only really started about 1000 B.C. in Central and South America, and had spread to the south-western United States about the beginning of the Christian era. At that time there were three alternative theories about the origin of maize, in the absence of any known wild ancestor. One possible source of the plant was a wild grass from Mexico and Guatemala, called teosinte; this had much of the appearance of maize but was plainly not closely enough related to maize, by botanical criteria, to have been a direct ancestor. Secondly, there was the South American wild grass, called tripsacum, whose claims were similar to those of teosinte and about as weak. Thirdly, and rather despairingly, it was suggested maize might have come from an extinct plant in South-East Asia.

MacNeish began with not much more than a feeling that agriculture was a necessity for the beginning of civilization, and that in the Americas agriculture meant maize. (As a European the author calls the plant *Zea mays* maize; Americans call it corn.) But no wild maize had ever been found, and maize, as we know it, is quite incapable of surviving under natural conditions because the ear is completely enclosed by the husks and the plant has no way of dispersing its seeds.

In 1945 MacNeish investigated some coastal sites in Tamaulipas, which is the first Mexican state across the border from Texas. He was officially searching for prehistoric cultural links between Central America and the south-eastern United States. He found two early cultures of food-gathering peoples who seemed to have taken the first steps towards agriculture, but there was no trace of any maize. The most important discovery was MacNeish's realization that cave sites in very dry areas could contain many stratified levels of deposits containing very large amounts of vegetable remains – remains not normally found on very early living sites but sometimes preserved by great aridity.

Two years later Herbert Dick started finding a great deal of vegetable matter preserved in Bat Cave, New Mexico, and by 1950 he had shown a steady succession of maize types starting at the bottom with very small cobs, only two or three centimetres long, and showing distinct evolutionary progress in several different parts of the plant as he got towards the top layers. Two important further steps were taken with this material: first, the lower layers, containing the tiny cobs, were dated to at least 3600 B.C. from charcoal found in them which was given one of the early radiocarbon datings; and secondly, samples of the maize were submitted to a botanist and plant-geneticist, Paul C. Mangelsdorf, Professor of Natural History at Harvard University.

The dating put the start of the cultivation of maize two thousand years earlier than anyone had supposed. But, even more important,

Mangelsdorf saw that the very earliest cobs were of two different types, a popcorn (the sort that explodes when heated) and a pod-corn. So he started crossing modern popcorns with modern pod-corns and continuously recrossing the hybrids back to popcorn in an attempt to reconstruct the original form of maize. He eventually produced a pod-popcorn plant which had two methods of dispersing its seeds in natural conditions, because some seeds were on the fragile tassels and the highest seeds on the ears were so high on the stalk that the husks did not cover them when the plant was ripe. This gave the archaeologist some idea of what to look for if he was intent on discovering prehistoric wild maize.

At almost the same time, in 1949, MacNeish himself found very early maize in a cave in Tamaulipas called La Perra. Again there was clear evolutionary development from the lowest layers up to the top, but the radiocarbon date of the lowest layer was only 2500 B.C. However, since the smallest cobs at La Perra were rather larger than the earliest cobs in Bat Cave, a later date was consistent, and even strengthened the theory that was now developing about the origins of maize. MacNeish, too, sent his samples to Mangelsdorf who found that they were closely related to a type of maize still found in Mexico, called Nal-Tel. The remains which MacNeish had found were, in Mangelsdorf's opinion, the ancestors of Nal-Tel maize, and the importance of this lay in the fact that Nal-Tel has been identified by botanists as one of the four ancient indigenous races of Maize in Mexico.

While working at La Perra, MacNeish also reconnoitred some caves in south-western Tamaulipas and he became convinced that earlier forms of maize, perhaps even prehistoric wild maize, would be found there. He teamed up quite deliberately with Mangelsdorf in these years in an attempt to find the origin of maize, believing that only the combination of botanist and archaeologist could solve the problem. However, the new caves, Romero's and Valenzuela's, proved at first disappointing. The oldest cobs were not much more than 2000 B.C. in date. They were definitely domesticated, and not wild, corn, but they were of a different type from the Nal-Tel ancestor at La Perra, and were more like the Bat Cave forms. All the evidence pointed to the domestication of maize as having taken place further to the south than Tamaulipas. But there was one extremely interesting finding – the discovery of remains of teosinte dating back to 1400–1800 B.C. Now teosinte does not grow in Tamaulipas at the present time so it was possible that the prehistoric farmers had deliberately grown teosinte in their fields of primitive maize in an effort to get better hybrid plants, a practice which has been recorded in modern times in western Mexico.

Other workers began producing information which helped in the search. Work in the Mexican states of Sonora and Chihuahua, both adjacent to the U.S.A. border, enabled Robert H. Lister to show that, although there were early examples of cultivated corn there, the domestication had almost certainly taken place further to the south. Then deep drillings in the Valley of Mexico by P. B. Sears and K. H. Clisby showed that there was a type of maize pollen at depths which corresponded to an Ice Age date of 80,000 years ago. This pollen was provisionally identified as coming from an unknown wild maize. It was certainly different from the pollen from domesticated, cultivable maizes in the upper layers. This implied, first, that the original wild maize had evolved in MesoAmerica, and, secondly, that domestication had taken place even further south again.

So in 1958 MacNeish moved his search far south, to find dry caves in Honduras and Guatemala. He found cultures of a pre-pottery type, but no corn. Then in 1959 he moved to the state of Chiapas, the most southerly division of Mexico, where he excavated in Santa Marta Cave. But pollen found there was considered too recent for the origins of maize to be placed so far south. The hunt must therefore concentrate in central Mexico, north of Chiapas, but south of the great central Valley of Mexico.

In what sort of area should the search for the first domesticated maize begin? Because maize likes plenty of water and has poor resistance to drought, Mangelsdorf favoured the humid jungly lowlands. MacNeish, however, argued for a highland area with low rainfall. There are several areas of the central Mexican highlands where the entire rainfall of the year is concentrated in the months from April to October, which is the growing season for maize. In the rest of the year these places are totally dry, virtually a desert, with cactus and similar 'xerophytic', drought-resistant, plants as the main vegetation. These perennials seem to avoid the riverside alluvial soils, presumably because they are too wet in the rainy season. This would give a presumptive wild maize a virtually unchallenged ecological niche, MacNeish argued. Though maize itself is not drought-resistant, its seeds can lie dormant and unharmed for many months, thus lasting out the dry season, with virtually no chance of unexpected rainfall starting germination at the wrong time of year and exposing young plants to frosts.

MacNeish settled on the Valley of Tehuacan as fulfilling his requirements. It offered not only the sort of environment which he thought might have favoured the growth of an original wild maize and the development of agriculture, but as it was dry and broken terrain it offered the possibility of finding dry cave sites which might contain deep deposits of stratified vegetable refuse from generations of living. Tehuacan is south-west of the main valley of Mexico, near the borders of Oaxaca where the Olmecs probably started the earliest of the high cultures of Mexico, and in the land ruled by the Mixtecs at the time of the European arrival.

In January 1960 he launched his first proper reconnaissance. But after looking at three dozen sites, guarded first of all by a lieutenant, then by a sergeant, and finally by a private before he was left to his own devices, the Tehuacan Valley was beginning to look disappointing. A Mexican friend had, however, enlisted the aid of local schoolchildren and sent round a questionnaire about caves they knew. This produced three further possibilities including a cave alleged by a Señorita Berta Martinez to be on a hill called Agujereado south of the town of Coxcatlan. MacNeish was taken for a long, hot and dusty walk through the cactus scrub and across the side of the mountain by the lady's brother, Hector, but the moment he saw the cave he thought it was promising. For six days from 21 January, with the help of Hector Martinez and a guide, Pablo Bolanos, MacNeish dug a trial excavation – a hole two metres square behind a large rock in the centre of the cave – rock-shelter is probably a better description than the word cave. Everything that came out of the hole was put through a mesh-screen so that they should miss nothing. Soon they were down to the period known as Classic in Central American archaeology, and below that they found remains of the period called Formative. Then came a sterile layer that had not been inhabited by anyone, and below again they came to dirt that was obviously a very thick layer left by people who had not even been able to make pottery (a pre-ceramic culture). MacNeish's own words describe what happened next:

On January 27 after lunch, Pablo, working well down in the preceramic stratum, recovered a tiny corn cob no more than an inch long. Only half-believing, I took his place at the bottom of the pit. After a short period of trowelling and cleaning away dirt with a paint brush, I uncovered two more tiny cobs. We held in our hands possible ancestors of modern domesticated corn.[1]

Radiocarbon dating showed the cobs to be older than 3600 B.C., the oldest corn cobs that had ever been found. Mangelsdorf confirmed that, botanically, they were ancestors of our maize. It seemed reasonably certain that the Tehuacan Valley was, at least, one of the places where maize had first been domesticated.

So the Tehuacan Valley Project was launched later in 1960, administered by the Peabody Foundation and supported by the National Science Foundation. Its purpose was, bluntly, 'To investigate the development of agriculture and the concomitant rise of civilization in MesoAmerica', and MacNeish hoped to find in that one valley a complete sequence of sites and strata which would give a picture of the lives, diets, habits and cultures of the inhabitants from before the beginnings of farming to the Spanish Conquest. There was to be a small field-team of archaeologists, but the 'experts', botanists, palaeontologists, ethnographers, zoologists and even malacologists, were to be associated from the start. Furthermore, they would all be brought out to Tehuacan at one stage or another; the publication of all the results and the implications of those results were also to be co-ordinated.

It was in January 1961 that the first archaeologists, including MacNeish himself and Frederick Peterson, arrived in the valley and set up their headquarters in a large rented house in downtown Tehuacan. The work of digging went on for the next three seasons. In all, sixteen pits were dug and a total of 156 occupational zones were found. Several caves and rock shelters were dug, and the final survey showed nearly four hundred sites in the Valley where archaeological materials could be found. But the original cave found by MacNeish on the advice of the Martinez family – renamed Coxcatlan cave proved the most valuable of all. It provided no fewer than twenty-eight different strata. By the end of their programme the expedition had collected a quarter of a million clay artefacts, 13,000 stone artefacts, 25,000 remains of domesticated plants, 80,000 remains of wild plants, 25,000 shells, 11,000 animal bones and 70 human burials. In particular they gathered from the five cave sites, preserved by the dryness, a total of 23,607 specimens of maize, of which nearly 13,000 were whole or almost complete cobs. But there were also specimens of roots, stalks, husks, leaves and other parts of the plant.

From all this material, the botanists, led by Mangelsdorf, were able to build up a complete evolutionary sequence of the development of maize from a wild type up to domesticated plants, which are clearly the ancestors of at least two of the present types of maize grown in Mexico. This implies that maize was domesticated independently in MesoAmerica from a wild plant which is now extinct. At least one of the places where this happened was the Tehuacan Valley.

The argument and the evidence goes broadly like this. The earliest cobs which were found in what MacNeish has named the El Riego and Coxcatlan Phases were small, and virtually uniform in size and characteristics, as might be expected from a well-established wild plant. Furthermore, these cobs show that at that stage the maize could drop and spread its seeds by the same sort of mechanisms as other wild grasses. The date of these cobs was given by radiocarbon dating of contemporary samples as between 5200 and 3400 B.C. (The recalibration of radiocarbon dates as suggested by the Stockholm Conference of 1969 would put these dates between six hundred and eight hundred years earlier, i.e. 6000 to 4000 B.C. The recalibration throws none of MacNeish's conclusions into doubt, but simply gives an earlier start and longer period of development to his general time-scale.) Finally the corn-cobs recovered from the following time-span, the Abejas Phase, are much more variable in size, including many large cobs, which implies that the maize was growing larger in the better conditions provided by cultivation. Other plants were also being cultivated in the Abejas Phase. MacNeish and Mangelsdorf conclude: 'This combination of circumstances leads to the conclusion – an almost inescapable one – that the earliest prehistoric corn from the Tehuacan caves is wild corn.'

But the wild corn did not disappear at once. In the Abejas Phase it provided 47 per cent of the samples, and occasional specimens were discovered in occupation layers dating as late as A.D. 250. Why did the wild corn finally become extinct? MacNeish answers this question:

We have for some years assumed that two principle factors may have been involved in the extinction of corn's ancestor. (1) The sites where wild corn grew in nature might well be among those chosen by man for his earliest cultivation. (2) Wild corn growing in sites not appropriated for cultivation but hybridizing with cultivated corn after the latter had lost some of its essential wild characteristics, would become less able to survive in the wild. Of these two causes of extinction the second may have been the more important. Corn is a wind-pollinated plant and its pollen may be carried many miles by the wind. It is virtually inevitable that any maize growing wild in the Valley would have hybridized at times with the cultivated maize in nearby fields which was producing pollen in profusion. Repeated contamination of the wild corn by cultivated corn could eventually have genetically 'swamped' the former out of existence. There is now good archaeological evidence in Tehuacan to suggest that both of these assumed causes of extinction were indeed operative. The alluvial terraces below San Marcos Cave, where wild corn may once have grown, now reveal the remains of a fairly elaborate system of irrigation, indicating that the natural habitat of wild corn was replaced by cultivated fields. Abundant evidence of hybridization between wild and cultivated corn is found in the prehistoric cobs. We have classified 252 cobs as possible first generation hybrids of wild corn with various cultivated types, and 464 cobs as backcrosses of first generation hybrids to the wild corn.[2]

The maize which MacNeish calls 'early cultivated corn' is exactly similar to the wild corn except that its cobs are clearly much larger. This type of plant became the most frequent in the Abejas Phase about 2500 B.C., but, before that phase had finished, a second new type of maize started to appear. MacNeish calls this 'early Tripsacoid' because the botanists showed that it was a hybrid of pure maize with the tripsacum grass, or with teosinte, both of which had earlier been considered as a possible ancestor for all maize. Now the problem is that tripsacum and teosinte do not grow in Tehuacan Valley, and apparently never did grow there, for the excavators found no trace of them by themselves. Therefore either Tehuacan 'early cultivated maize' was taken out of the Valley and grown perhaps in the neighbouring state of Guerrero, where both tripsacum and teosinte are natives of the Balsas River area, and then, after natural hybridization, the improved plants were brought back to Tehuacan, or there were other places in Mexico where maize had been domesticated independently and where hybridization had occurred. MacNeish seems to favour the latter possibility.

The 'early tripsacoid' maize dominated the crop by about 1000 B.C. but at about that same time there come the first appearances of the direct ancestors of the Nal-Tel race of maize, which is still found in Mexico. Or, putting it another way, the type of maize grown about 1000 B.C. contains samples like those found by MacNeish in Tamaulipas. After that come other new varieties, and from the occupation layers representing the time of the Spanish Conquest the excavators recovered specimens of maize of several different types which are perfectly well known to present-day Mexican farmers. A complete and satisfactory evolutionary sequence for the domestication and development of maize had been found.

In one sense, therefore, MacNeish had successfully completed his search for the origins of maize. He had proved it came from the now extinct wild maize, and he had archaeological evidence of the development of modern forms of the plant. He concludes:

The only real changes in more than 5000 years of evolution under domestication have been changes in the size of the parts and in productiveness. The importance of these changes to the rise of the American cultures and civilizations would be difficult to overestimate. There is more foodstuff in a single grain of some modern varieties of corn than there was in an entire ear of the Tehuacan wild corn. A wild grass with tiny ears – a species scarcely more promising as a food plant than some of the weedy grasses of our gardens and lawns – has, through a combination of circumstances, many of them perhaps fortuitous, evolved into the most productive of the cereals, becoming the basic food plant not only of the pre-Columbian cultures and civilizations of this hemisphere but also of the majority of modern ones, including our own.[3]

Nearly ten years later, however, there are warnings about one of these conclusions. Modern botanists point to the vulnerability of the maize plant of modern times because man has forced it to evolve until it cannot live in nature without cultivation because it cannot spread its own seeds. H. Garrison Wilkes, a botanist of the University of Massachusetts, has pointed out that wild teosinte and similar grasses which provided some of the genetic materia of the modern maize by hybridizing with the 'early cultivated' maize are threatened with extinction by exactly that combination of circumstances which MacNeish suggests extinguished wild maize. The genetic basis of modern maize is therefore becoming dangerously narrow, and the total extinction of teosinte would ruin any chance of broadening it again.

But with the Tehuacan Valley excavations, MacNeish discovered much more than just the origins of maize. With the help of other botanists, specializing in different types of food plants, and using the same sort of arguments, they were able to show that the chilli-pepper, the avocado pear, the squash plant called *Cucurbita mixta*, the white and black sapotes and various sorts of beans were all cultivated at very early stages in the Tehuacan Valley. The chilli-pepper and the avocado may have been cultivated even before maize – that is, back in the El Riego Phase as early as 6000 B.C. Others of these plants first appear in the Tehuacan Valley as cultivated plants with no earlier evidence of their having been native to the region. To MacNeish this suggests that the origins of agriculture must have been a complex process with different plants being first cultivated in different centres of MesoAmerica.

Using a different line of approach, the calculation of the proportions of the diet of prehistoric man from the various different sources, it was also shown that total dependence on cultivated food took a very long time to develop. This exactly parallels the discoveries at Deh Luran. The Tehuacan calculations show that if the very first cultivation began about 7000 B.C. we can say that at that date about 60 per cent of the inhabitants' food came from wild animals and 40 per cent from wild plants. Thereafter, the contribution of animals decreased and the proportion from wild plants actually increased to such an extent that wild plants provided 52 per cent of the total diet at the end of the Coxcatlan Phase, say about 3500 B.C. At that time cultivated plants provided only about 15 per cent of the total diet. The first domesticated animals in the Tehuacan Valley only appear about 3000 B.C., and even by the Spanish Conquest they were providing less than 5 per cent of the total food intake and wild animals were still providing 10 per cent. Agricultural plants only started providing more than half the food by about 1000 B.C., and wild plants were still providing about 8 per cent in the final phase, the Venta Salada Phase, just before the Europeans arrived.

Before any cultivated plants were available, the inhabitants of the Tehuacan Valley appear to have lived largely on meat; it provided 70 per cent of their food intake, with wild plants providing the rest. In this earliest phase, the Ajuereado Phase, the people were at the cultural stage of 'big game hunters' or 'mammoth hunters', a cultural phase which has gained its name from discoveries in other parts of America. It seems, however, that the inhabitants of Tehuacan were very much less grandiose than 'mammoth hunters'. Although bones of animals now extinct in Mexico have been found at their sites in Tehuacan, the vast majority of their food clearly came from very small game such as rabbits and rats.

The actual diet of the inhabitants of the Tehuacan Valley is known not only from careful, but necessarily rather imprecise, calculations of the proportions of food remains in their living floors, but also from the analysis of their excrement. A considerable number of what the archaeologist terms 'coprolites' were recovered from the various caves – these are in fact dried stools. They were 'reconstituted' by Eric O. Callen, the man who first originated this fascinating though, he admits, often very smelly type of study, by soaking them in a solution of trisodium phosphate. Allowing for the fact that some foods, especially prepared and cooked foods, are virtually unidentifiable in faeces, the broad results he obtained agreed remarkably well with the over-all analyses of diet contents from cave remains, despite disagreements between Callen and MacNeish on some points. Callen also made one unusual and unique contribution to the Tehuacan results, in that he believes he has identified, in the later periods, a noticeable difference between a 'city' diet of beans, pineapple and maguey with a little meat, and a rougher, 'cave' diet, which dates back in origins to earlier phases. The implication is that people were living in towns and villages on a diet consisting entirely of cultivated foods, and visiting the country caves occasionally when they either changed their diet or met country cousins living on a different diet. Callen was also able to identify from certain coprolites that the human producer of the stool had been living on a 'wet season' diet or a 'dry season' diet according as the remains of food were those of plants available in the wet or dry seasons of the year.

The evidence of the coprolites supported the more conventional archaeological evidence, the evidence of stone tools, and pottery, the first findings of grinding-stones and pestles, the discoveries of early textiles and basketwork, and the evidence of seventy graves, to enable MacNeish and the sixty associated scientists to build up a picture not only of early agriculture, but also of a whole series of developing human societies.

There were probably only between twelve and twenty inhabitants of the whole valley of Tehuacan in the 'mammoth hunting' or Ajuereado Phase. Technically they were much more hunter-gatherers than 'mammoth hunters', for fully 30 per cent of their diet was gathered from wild plants. The pollen and animal bones imply that the climate was just a little wetter and cooler than at present, but a widely held view that there were major climatic changes in the Americas about 7000 B.C. is not supported. Up to about this date of 7000 B.C. the twenty or so inhabitants of the valley were split into three 'microbands', small family groups of four to eight people. They moved in regular annual rotation from a wet-season camp to a dry-season camp and an autumn (fall) camp.

Between 7000 B.C. and 5000 B.C. – up to what MacNeish calls the early Coxcatlan Phase (dates uncorrected for radiocarbon tree-ring calibration) – the population of the valley quadrupled to perhaps nearly one hundred people. These people remained in 'microbands' with separate small camps for the autumn and the dry season. But once a year they seem to have joined up into three large 'macrobands' at regular spring camp sites. The implication is that this gathering together was to use the lush vegetation brought by the rains. They were certainly eating plants which were later cultivated, such as squash, chilli-pepper and possibly maize, and the obvious suggestion is that this was the time when the idea of cultivation arose. MacNeish says, 'I would guess that this was the period when they finally conceived the idea that if you drop a seed in the ground a plant comes up.' Certainly these people of the El Riego and Coxcatlan Phases were beginning to develop relatively elaborate burials which indicate that they may have held quite complex beliefs which they expressed in ceremonies. There were cremations and suggestions of human sacrifices in the graves, where the bodies were carefully wrapped in blankets with netting and basketwork. 'Is it not possible that the ceremonialism that is so characteristic of the later Mexican periods began at this time?' MacNeish asks.

By 3400 B.C. the Coxcatlan Phase had developed into the Abejas Phase. We have seen that this marks the time when the cultivation of maize can definitely be established. Chilli-pepper, squash and avocado pear were probably cultivated slightly earlier than corn; beans and gourds came soon afterwards. The population of the valley must have been nearly two hundred by this date and the macrobands lived in one main camp for much of the year, separating into hunting, trapping and gathering microbands probably only during the dry season when food was at its most scarce.

MacNeish estimates that from 3000 B.C. to 1500 B.C. the population of the valley increased to about eight hundred, still living mostly in semi-permanent villages, with occasional small camps for microbands who went out hunting or planting. Agriculture was steadily increasing and the number of different domesticated crops also increased. Then, in the centuries around 1000 B.C., permanent villages, with temples, appear as the steadily increasing agriculture allowed the valley to support a few thousand people. Irrigation may also have begun about this time. There are plenty of other archaeologically known sites in Central America which can be compared with this stage of Tehuacan history, and it must be at about this time that the MesoAmerican world split into the two distinct cultural units, one based on the lowlands with a slash-and-burn jungle agriculture which produced the Maya civilization, while the other, Tehuacan-type, highland culture developed agriculture based on irrigation. From this stage certainly developed the civilizations found by the Europeans when they conquered Mexico.

The Tehuacan Valley has, then, provided a continuous sequence of human social evolution as well as showing the evolution of domesticated crops under the influence of cultivation and agriculture. MacNeish sums this up, and asks further questions in the final words of volume one of his *The Prehistory of the Tehuacan Valley*.

One gets the impression that the development of civilisation and more effective food production in MesoAmerica is not due to a single evolution of developmental stages of culture and subsistence, but that a series of concomitant developments of rather different ecological zones are interacting with and interstimulating one another in such a manner as to bring about cultural development and increasingly effective food production. In fact is not this sort of symbiotic development of agriculture and culture one of the causative processes that leads to effective food production and civilisation in MesoAmerica? Or, to put it another way, was not the development of effective food production and the concomitant cultural development in MesoAmerica in no small part owing to the fact that there were contiguous environments or ecological zones that were exploited agriculturally in different ways and evolved through different cultural stages, and their geographical closeness allowed the varied subsistence developments to interact and interstimulate one another? Further, is not this symbiotic process which developed effective food production, and civilisation, in MesoAmerica, the same sort of process which must have occurred in other areas of primary civilisations such as the Near East, Peru, or even China? If so, can we make universal and unilinear generalisations, other than the broad kind of statement that food-gathering preceded food-production or that cultures developed from simple to complex? In fact must not our generalisations about the development of agriculture and the rise of civilisation be of the variety that Stewart has called 'multilinear evolutions'? Is it not time even with our meagre data, to think in these terms?[4]

The work at Tehuacan, although it pushed back the date of the beginnings of cultivation in the New World to a time almost contemporaneous with the same development in the Middle East, did not change the time-scale for man's existence in the New World, though it involved accepting a very rapid development indeed from coming across the Bering Straits in 15,000 B.C. to developing agriculture only eight thousand years later. But in fact the latest work all over the western hemisphere has greatly changed the position in the last five years.

MacNeish himself started a further project, very similar to the Tehuacan project, in Ayacucho in the Peruvian highlands. The discoveries there, where Flea Cave has proved one of the most exciting sites, have shown man to have been present 22,000 years ago. There have been controversial finds at Calico in the Mojave Desert which could be equally old, and definite datings for material found in Tlapacoya and Hueyatlaco in Mexico at 23,000 and 22,000 years old. Likewise, bone tools from a site called Old Crow in the Canadian Yukon have been radiocarbon dated to 23,000 to 28,000 years old. A date of 38,000 years has been estimated for some flake tools found at Lewisville in Texas. MacNeish has suggested that at least three of the different cultural traditions found in these very early sites may have come from Asiatic sources, and been brought across to America as a culture. This would place the arrival of man in the New World somewhere around 50,000 B.C., but this is admittedly in the realm of speculation.

It is worth adding that the most recent report to hand gives evidence that common beans and lima beans were cultivated in Peru as far back as 8500 B.C., ten thousand years ago. The report comes from Lawrence Kaplan, of the Biology Department of the University of Massachusetts, with Thomas F. Lynch and C. E. Smith. Kaplan was responsible for analysing the bean remains at Tehuacan and the earliest examples of cultivated beans he found there dated back to 7000 B.C.

The beginnings of man and the beginnings of his agriculture in the New World still remain, therefore, an open question, but the Tehuacan Valley Project provided, for the first time, a real platform for discussion, and a set of facts which completely upset all previous notions about these fascinating problems.

1. P. Mangelsdorf *et al.* 'The Domestication of Corn'.
2. ibid.
3. ibid.
4. R. MacNeish in the final chapter of D. Byers ed., *The Prehistory of the Tehuacan Valley*, vol. 1.

10

Traces, and How to Find Them

Throughout the history of archaeology, the archaeologist has relied on his eyes. He deals with what he can see; he digs where he can see a likely place or some telltale sign. Leonard Woolley illustrated the point well:

The archaeologist, in fact, has to keep his eyes open for evidence of all sorts. At Carchemish, in North Syria, my old Greek foreman, Gregori, an experienced digger if ever there was one, and I, completely mystified our Turkish inspector. We told him we were going to excavate a cemetery, and as we had not previously found graves he was duly interested, and asked to be shown the spot. We took him outside the earth ramparts of the old, city to a ploughed field by the river bank, lying fallow that year, and pointing to the fragments of pottery which strewed the ground, explained that these constituted good evidence for the existence of a graveyard. Then Gregori and I, consulting together, started to make piles of stones marking the position of individual graves. This was too much for Fuad Beg, who protested that we were bluffing him; I betted that we should find a grave under every pile and no graves at all except where we had put a mark; he took the bet and lost it, and spent a month wondering why. It was a really simple case of deduction. The river bank was of hard gravel, the made soil overlying it very shallow, and disturbed to the depth of only about three inches by the feeble Arab plough; the field, being fallow, was covered with sparse growth, for the most part shallow-rooted, but with a mixture of sturdier weeds of a sort whose roots go deeply down; if one looked carefully it became manifest that these weeds were sometimes single, but often in clumps of four or five plants, but a clump never measured more than six feet across; at some time or another the gravel subsoil had been broken up, so that the plant roots could penetrate it, and it had been broken up in patches which would be just the right size for graves; the broken pottery on the surface represented either shallow burials or, more probably, offerings placed above the graves at ground level, and every deep-growing weed or group of weeds meant a grave-shaft. The deduction proved correct.[1]

The archaeologist's eye coupled to a trained and experienced mind was obviously a most powerful tool for finding things that are hidden. But the laboratory scientist has developed even more powerful tools than eyes; there are families of devices which can 'see through' several feet of earth; there are other tools – the microscope is the most obvious – which can 'see' things so small that the naked eye could never tell they were there. The application of these various scientific 'seeing' devices is another area in which science has improved the quality and range of data that can be recovered from the archaeological record. In turn this has changed the techniques with which archaeologists dig.

These scientific methods of detecting traces of things the eye cannot see fall into two categories. First there are the scanning systems which tell the archaeologist where to dig, or even whether it's worth digging at all. These systems present the broad view, they show things that the archaeologist on the ground cannot see because the 'spade's-eye view' is too limited. The second category is the extremely detailed view – the observation of minute traces of materials such as chemical elements which may be present as only a few parts per million in something the archaeologist has dug up. Inevitably this sort of analysis element by element, almost atom by atom, requires expensive equipment and highly sophisticated laboratory techniques. Yet it presents an ability to read out from the archaeological record more than the eye can see in it, and when it is applied it vastly increases the amount of information that can be recovered from the archaeologist's activities. Alternatively, the use of these techniques can give a resoundingly negative answer to some hypothesis which is otherwise plausible.

Looking first at the broad-scanner techniques the obvious example that leaps to mind is aerial photography. This began a very long time ago, even before aeroplanes were invented. It was in 1891 that the first suggestion that vertical photographs taken from a great height would reveal archaeological features not observable from the ground was put into practice. Balloons were used and photographs were taken over Calcutta for the Surveyor-General of India. Largely because of lack of official enthusiasm and mishandling of the project, they were failures. The first major series of aerial photos of archaeological sites was produced by Dr Theodor Weigand in 1920 as a result of his work when attached to the German armed forces operating in southern Palestine and the Sinai Desert during the First World War. They definitely showed that there were several abandoned cities under the surface of the desert and the whole plan of some of these cities, even down to the vineyards, was clearly shown. Wider interest in the technique of aerial photography was aroused three years later when it was found that pictures taken by a Royal Air Force plane, a couple of years earlier in the very dry summer of 1921, clearly showed the full extent, for the first time, of the great avenue leading to Stonehenge, the most imposing of all the megalithic monuments of Europe, on the downlands of south-west England.

There have been many uses of aerial photographs in the service of archaeology since then, but despite the dramatic results of air surveys and the clear pictures they have presented, there has been a strange unwillingness to make the partnership between the two into a regular marriage. Perhaps it has been the perennial lack of funds that always affects the archaeologist; perhaps it is that the aerial view is always too much out of scale with the archaeologist's powers on the ground below – the air photo covers a couple of square miles of ground while the archaeologist knows that he cannot excavate more than a few hundred square yards. So we have in 1971 this statement, 'Over this rather extensive period of time the use of black and white aerial photography has not been as extensive as one might have anticipated'; the author is Dr George J. Gumerman, Director of the Archeological Survey and Assistant Professor of Anthropology at Prescott College, Arizona. He insists that it is only by seeing aerial black-

and-white photography as one of a battery of techniques that bring new powers of 'remote sensing' to the archaeologist, that the proper use of the techniques will be defined and really valuable results will emerge. 'It is only recently, however, that archaeologists have placed methodological emphasis on aerial panchromatic photography and other sensors as a means of prediction rather than simply as a means for recording and interpretation,' Gumerman writes. In addition to ordinary photography, both black-and-white and colour, there is infra-red photography which produces pictures in black-and-white and colour, and is particularly informative about soil types and conditions. Radar can be used, too, as a remote sensor producing interpretable 'pictures' for the archaeologist.

The importance of using aerial photos for prediction, rather than just for mapping, was shown in a study of the Estancia Valley in New Mexico, when the photographs showed the beach terraces of lakes that had dried up thousands of years before. In particular it could be seen that a peninsula of land with water on three sides would have made an admirable place for Palaeo-Indian hunters to drive and trap mammoths and extinct forms of bison which were their usual prey. Ground surveys then duly found several early sites of man's occupation dating back to Folsom times.

Many Europeans would feel that Gumerman is relying too much on American experience in his statements quoted above. In Britain particularly aerial photography has been extensively used in aid of archaeology, probably, indeed, the technique is particularly suited to such a crowded island with such a dense pattern of historical occupation. The peak of activity in aerial surveying of the British Isles is now probably over, but the result of all the work is a superb National Collection of aerial photographs, which is treated as part of the working equipment of the archaeological sections of the Department of the Environment. This deposit of pictures can be used for rescue work, to see what archaeological 'properties' are likely to be affected by major developments such as new motorways across the countryside. It can also be used for more typically scholarly work, such as the study of the different distributions of field and cultivation systems across the country. Aerial photography has also been used extensively to study the 'centuriation', or land-measuring and surveying systems of Roman times, around the entire shore of the Mediterranean.

The next major development expected in aerial photography is the increased use of infra-red photography. This should show up the markings of ancient works underneath crops which do not register clearly in black-and-white pictures. Because of the different heat capacities of different types of earth, or of disturbances within one type of earth, it should also reveal other features of the past. Infra-red photographs from satellites have already shown how clearly this technique can register differences in underground water supplies or differences between healthy and infected crops. It may well throw up unexpected archaeological finds in the course of time.

Nevertheless, much more impact has been made on archaeology by a succession of devices developed since 1946 which use electrical, electronic and magnetic means to scan likely 'digging areas'. These techniques tell the archaeologist exactly where his digging is likely to be most fruitful. The devices are of two families, those depending on measurements of the electrical resistivity of the earth, and magnetometers. But both families work, in practice, because they can 'sense' down through the earth not only for 'hidden treasure' but for disturbances in the soil, places where something has been dug or built in times past and where, as a result, there are variations in the otherwise steady patterns of electrical current or magnetic field.

The search for Sybaris was an outstanding example of the magnetometer in action. Sybaris was one of the classical Greek colonies in southern Italy, and it turned out so prosperous that its name lives with us today in the word 'sybarite', synonym for the affluent, lazy yet pleasure-loving man. But the ancient Sybarites met their come-uppance at the hands of the neighbour city Crotona in 510 B.C. and their opulent town was burned and levelled to the ground. 2,500 years later all that was known of it was that it must be on the delta-plain of a river that flowed into the Gulf of Taranto, an area still called the plain of Sybaris. But this plain covers forty square miles and it has a history of being racked by earthquakes. The very first survey by a University of Pennsylvania Museum team in the early 1960s showed that one earthquake had lowered a section of the plain by fifteen feet and allowed the sea to rush in and form a lagoon; this lagoon, however, had silted up, been flooded, silted up again, and Sybaris, or its remains, could lie anywhere between twelve and twenty feet down. At this stage the Museum Director, Dr Froelich Rainey, brought a proton-magnetometer to the scene which succeeded in discovering the ruins of the Roman city which had been built centuries later than the original Sybaris. The Museum's Applied Science Center for Archeology was meanwhile developing the much more sensitive caesium-magnetometer. In 1967 Dr Rainey brought the new machine to the plain of Sybaris for its first major test in the field. He was able to survey ten acres a day penetrating down to twenty feet below the surface. In due course the instrument found magnetic disturbances in certain places at depths of about twenty feet, and test borings in the places where the disturbances had been found brought up ancient roof-tiles and potsherds. Everything was wet because the water level is only three feet below the surface at this point, but small trenches confirmed that the foundations of buildings were down there, and Sybaris had been found. Further work with the caesium-magnetometer showed that ancient riverbeds lay both north and south of these buildings, confirming the ancient description of Sybaris as lying between two rivers.

Since then the caesium-magnetometer has scored several more important successes. The year after Sybaris the Associate Director of M.A.S.C.A., Dr Elizabeth K. Ralph, was able to map the main outlines of the Greek city of Elis, about forty-five miles from Olympia, although most of the remains were hidden under crop-fields, by using a new and more portable version of the machine. By the following year the instrument had been developed to give audio-signals corresponding to its readings of underground variations in magnetic fields, and this, combined with a tape-recorder carried by the surveying party, enabled even more rapid surveys of large areas to be made. It was taken by Dr Ralph to San Lorenzo, in the southern part of the Mexican state of Veracruz, where there is a very large Olmec site covering more than six square kilometres of mesa, but almost entirely buried. It has been dated 1200–900 B.C. Digging in the places indicated by the magnetometer, Dr Michael Coe of Yale University uncovered the finest example of an Olmec carved human head yet found and five other significant finds including an Olmec workshop.

The first magnetometer was a proton-magnetometer, and the first suggestion that such a machine might be made to help archaeologists came from a nuclear-physicist at Cambridge University, Dr J. C. Belshe, in 1956. This was only two years after the discovery of the rather obscure phenomenon of nuclear physics, on which all the subsequent machines are based, had first been applied to geophysical work. The basic discovery was that the nucleus of a hydrogen atom – a proton – seems to spin around. When a child's top is spun in the field of the earth's gravity it does not stay steady and upright, it gyrates. The proton at the centre of a hydrogen atom behaves similarly in the earth's magnetic field. The stronger the magnetic field, the faster a proton in it will gyrate. If there are enough hydrogen atoms present these gyrations can be sensed as an alternating current in a near-by electric circuit, and the frequency of the gyrations appears as the frequency of the alternating voltage in the circuit. The principle is simple, even if the circuits and electronics necessary to convert this phenomenon into signals appreciable by a human being are fairly complex. Because the principles are simple the basic sensing device of a proton-magnetometer is simple – it is nothing more than a bottle of water, which contains a great many hydrogen atoms according to the well-known formula for water H_2O. About a thousand coils of thin conducting wire are wound around the bottle. Thus a proton-magnetometer has a very light, portable and easily manoeuvrable sensor head, from which a light electric cable can lead back to the electronics and meters as much as a hundred yards away. It is extremely sensitive to the variations in the magnetic field caused by disturbances of the earth from former ditches or pits or the foundations of long-buried buildings.

Belshe's idea was one of the first devices to be taken up and developed by the then recently founded Research Laboratory for Archaeology in Oxford under Dr E. T. Hall. By March 1958 the first instrument was ready for a field trial and it was rushed out to the place where a new road was about to be built through an area which was known to have been the site of many Roman pottery kilns near the Romano-British settlement of Durobrivae. Because these kilns produced a type of pottery called 'colour coated ware', which was important to British archaeologists in building up their picture of the country during the times of Roman occupation, it was also important to locate any undiscovered kilns that lay on the route of the new road and would soon be lost for ever. Martin Aitken was the man who tried out the new device and he wrote, 'The survey was highly successful and the importance of the proton magnetometer for archaeology was immediately established.'

Since then the proton-magnetometer has been used successfully in many parts of the world on literally hundreds of archaeological sites, though probably its most notable achievements have been in surveying Iron Age hillforts and the like in Britain. A closely related instrument, a soil anomaly detector commonly called a 'banjo', has been in the public eye through its use in the last few years in surveying the large area inside the hillfort of South Cadbury in Somerset, where there has been much excavation in search of the historical realities behind the legend of King Arthur. The excavators did not find the Camelot that Hollywood portrayed, but they revealed a great deal about the Dark Ages when Rome withdrew and left the Romanized inhabitants of Britain to deal with the Saxon invaders.

The other family of electronic devices which the archaeologist uses for 'seeing' through the soil beneath his feet are those based on measuring the electrical resistivity of the ground. Once again these techniques were first developed by the geophysicists and taken over from them for archaeological prospecting. In this technique two electrodes are driven into the ground some distance apart and a potential difference is applied across the two. Two further electrodes are then used, in between the first two, to measure whether there are any irregularities in the pattern of potential difference in the electrical field set up between the two outer electrodes. Any disturbances in the earth due to the past activities of man will show up as disturbances in the electrical potential difference patterns.

This technique of resistivity measurements is particularly good at demonstrating underground buildings or the foundations of walls. It had originally the disadvantage of a lack of mobility compared with the magnetometers and it was at first necessary to operate the four electrodes in straight lines. Nevertheless even in its earliest forms resistivity surveying proved particularly successful in central Italy where the Lerici Foundation used the technique to locate more than ten thousand underground graves and tomb-chambers, mostly from the Etruscan civilization. Many of these Etruscan tombs had been plundered by tomb-robbers centuries ago, but the demands of the antique-market and the activities of twentieth-century tomb-robbers place them all at great risk. Even if they cannot all be excavated, they can at least be protected once their position is known to an official body.

For some years resistivity surveying took second place to the magnetometer unless it was known that buildings or foundations would be encountered (as in the Winchester excavations). But very recently there have been technical developments in the instruments for resistivity work which enable the archaeologists using this technique to get away from the former disadvantages of lack of mobility and straight-line working. Many of these developments have been pioneered by a team at the Research Laboratory of the Ancient Monuments Department of the British Department of the Environment. They include the production of a device rather like an electrified four-legged table, which is lightweight and highly mobile and in which the four metal legs can be used as the four electrodes of the early resistivity systems. This same laboratory has also introduced a new form of magnetometer, the flux-gate magnetometer, which overrides many of the problems faced by the earlier devices by using two separate magnetometers at the ends of a short vertical pole and recording only the differences between the two readings. This device is now being attached to a digital coder so that the results obtained by the archaeologist or scientist walking across the site on a series of grid-lines can be fed directly into a computer. It is hoped soon to record the precise position of the walking archaeologist at all times by a radio-location device, so that complete plots of the site can be drawn out by the computer automatically. Surveys done by the team from this laboratory have been used at many important sites all over the British Isles recently, notably at the Iron Age settlement at Gussage St Michael where recent excavation has turned up extraordinarily detailed evidence of early metalworking and casting techniques. The surveys are now much more than just indications to the archaeologist as to where to dig – they are accepted, especially at sites of secondary importance, as part of the formal archaeological record. They are also extensively used where sites are coming under development for housing or similar purposes; a good recent example of this is the record made by magnetometer survey of an Iron Age camp site and cattle-enclosure at Bauxbury near Andover, where a new housing estate is being built.

But if the magnetometers and resistivity surveys can find lost cities or trace a lost ditch for half a mile through a tangle of buildings and ruins of later ages, as was done at the great excavation of Roman Verulamium, now St Albans in Hertfordshire, the science laboratory has probably even more to offer the archaeologist when he is faced with a tricky problem of identification, when he asks the question 'Where did this come from?' Of all the scientific aids to archaeology this is the field in which the most rapid progress is being made at the present moment.

This development began in 1923 when H. H. Thomas, a geologist, showed that the famous 'Blue Stones' of Stonehenge, forming a separate small circle of their own inside the giant trilithons which dominate the skyline of the Downs, could only have come from the Preseli Mountains in South Wales. He had made a detailed analysis of the stones by the thin-section technique and showed the exact proportions of the main minerals in them – in other words, a petrological analysis. With this he was able to tie down the origin of the stones to one particular area of the Preseli Mountains, Carn Meini – and the Preseli Mountains are a very small range of very small mountains in Pembrokeshire. His work has never been seriously questioned since, but he left it to the archaeologists to say why anyone had transported enormous blocks of stone for a distance of 140 miles as the crow flies, which must have involved a far longer journey by any mode of transport Neolithic Man could have provided. Still less did he say how this transport could have been done, and the archaeologists have devoted much time to speculating or even experimenting with moving mock-prehistoric rafts up West Country rivers carrying enormous loads of unwanted stone. The irony of this story may well lie in the suggestion made within the last two years that prehistoric men never transported the stones at all; it is argued that there was one tongue of glacier in the Ice Ages which came down over Ireland and swept into south-west England in an easterly direction, crossing south Wales as it did so, picking up boulders from Preseli and depositing them as it retreated on Salisbury Plain, in the area that was later to be the neighbourhood of Stonehenge. The builders of Stonehenge then used the great blocks of Welsh mountain stone that they found lying about on the surface. Whether or not this theory about the ice-transport of the Blue Stones of Stonehenge stands the test of time and further critical examination, the whole story shows both the strength and the weakness of chemical analysis from the archaeological points of view. Broadly speaking, chemical analysis can establish the origin of things, artefacts or natural objects, in a way, and with an accuracy, that no other technique can achieve. But these discoveries by solving one archaeological problem may simply raise others, and, what is even more difficult, newer and more precise scientific techniques may reverse the conclusions drawn from earlier studies. The development of scientific techniques can be very rapid.

Thus as recently as 1969 Professor F. W. Shotton, a geologist from Birmingham University, wrote: 'No one has yet found micro-characteristics which will enable us to separate the flints of Yorkshire from those of Lincolnshire, Suffolk, Kent, Sussex or Antrim, still less to be more precise in location, so that thin-section work in this direction is unrewarding.' Yet in 1972, in the journal *Archaeometry*, two groups, one based on Bradford in Yorkshire and the other on Leiden and Delft in Holland, claimed that by the use of neutron activation analysis they believed they had found a system which could, in some cases at least, distinguish the products of different prehistoric flint-mines. The British Museum Research Laboratory also has a group making important progress along the same lines.

Similarly in 1947 Jacquetta Hawkes could write about one of the earliest attempts to apply advanced chemical analysis techniques to archaeologically discovered artefacts: 'It was the spectroscopic examination of the blue faience beads of the Wessex culture that made it perfectly certain that they were actual imports from Egypt.'[2] She is here referring to the gay little blue beads which were used at that time to date so much of British prehistory (see Chapter 2).

Faience beads are the earliest form of manmade jewellery. They were first made in the Mesopotamian civilizations long before 3000 B.C., and an Egyptian faience industry began before the dynastic period. In essence these beads have a core of primitive glass, or pre-glass, made by powdering quartz and then heating the grains so that they coagulate into the desired shape. A glaze is then applied and this glaze is usually blue-coloured by the addition of small amounts of copper. The first important analysis of the manufacture of faience in prehistoric times was made as long ago as 1927 by Sir Flinders Petrie as part of his great excavation of Tell-el-Amarna in Egypt. Sir Mortimer Wheeler in 1953 discovered that India had developed her own faience industry, and the techniques of manufacturing faience are known to have spread to Crete and later to Greece in the early part of the second millennium.

The faience beads found in British graves were first subjected to serious study in 1927, when all the 250 examples then known were classified and studied by Beck and Stone. Their final conclusion was that the beads must date, in their British context, from about 1400 B.C., which was the middle of the Bronze Age in these islands. They could find no evidence of the beads being manufactured in Britain, and very little evidence of similar beads being found in Europe. The beads were not very similar to Egyptian faience workmanship, but they could only conclude that they must have been imported from Egypt or Palestine. In the course of the next fifteen years further beads were found in western Europe, notably near Lake Constance in Switzerland and as far afield as the Scilly Isles. Since all the beads were found in graves reasonably near sea coasts or rivers it was postulated that trade routes had been built up by the Mycenean Greeks by sea and by the great European river routes, and they had brought the beads as trade goods from Egypt. Nothing else seemed to be able to account for the sudden appearance of these objects in the British archaeological record and their equally sudden disappearance after a few centuries. It was really not possible to imagine the comparatively sophisticated technology of faience-manufacture (which is still carried on to this day in the Middle East) being developed independently by the Bronze Age barbarians of Britain or Europe. It was at this time that Thomas performed a spectroscopic analysis of sixty of the British beads and could find no difference between them and Egyptian beads in their chemical composition. He concluded that the only sure guide to the identity of the manufacturers of the beads was purely archaeological evidence. It was on this study that Jacquetta Hawkes based her statements of 1947.

Thomas had noted, however, that the British beads seemed to have a rather higher tin content than other beads. Also, the beads from Bohemia and Moravia that he had examined at the same time seemed rather higher in cobalt than either Egyptian or British beads. These facts, which seemed unimportant before the Second World War, assumed greater significance when the general trend of opinion began to swing towards giving northern European developments a longer chronology. By 1968 Colin Renfrew was arguing that the very rich 'Wessex' culture of Bronze Age Britain was earlier than the Mycenaean civilization of Greece. With the more powerful methods of statistical analysis that had been developed with the coming of computers Newton and Renfrew re-examined Thomas's figures, and found that there were statistically significant differences between English, Scottish and Egyptian beads in respect of the aluminium, magnesium and tin contents of the faience. They emphasized the differences in shape between British and Egyptian beads, and pointed out that, apart from the beads, there was very little evidence to link the Mediterranean with central Europe in Bronze Age times, and even less to link central Europe with Britain. But they ended by pleading for a 'sharper tool' to discriminate between beads made in different areas.

In 1972 it was claimed that this 'sharper tool' had been found in the shape of 'neutron activation analysis'. Aspinall, Warren and Crummett from Bradford University with R. G. Newton of the Glass Industry Research Association looked at the faience beads yet again using this new technique. Thomas had provided them with the majority of the beads he had tested as recently as 1951; there were Egyptian and Cretan and Maltese beads, as well as central European examples to compare with beads found in England and Scotland. The beads were sent to the Herald atomic reactor at Aldermaston to be bombarded with neutrons, and then examined for radioactivity. The results clearly showed that the amount of tin in the British beads distinguished them quite clearly from central European and Mediterranean and Egyptian beads. The elements scandium and caesium were also clearly different in British beads. Egyptian beads, for their part, showed clearly higher levels of gold and silver than any other group. The central European beads, as Thomas had spotted, were much higher than any of the other groups in cobalt. The conclusion is clear: the chemical contents and the differences in shape imply that the faience beads found in British graves did not come from Egypt, and were quite possibly made in Britain. Likewise the beads found in central Europe did not come from the Mediterranean, and were quite possibly made locally.

Much larger studies on faience beads lent by British and Czechoslovak museums are now being set up to attempt to come to firmer conclusions by using a larger number of samples. But the 'Mycenaean trade routes' through Bronze Age Europe have become much fainter, perhaps they have even disappeared.

Neutron activation analysis involves putting the samples into an atomic reactor (usually) and bombarding them with neutrons, which makes them highly radioactive, since many atoms have been changed into their unstable radioactive isotopes. When they are removed from the reactor they give off this radioactivity and it has been found that each type of atom radiates at very precise wavelengths. By measuring the exact wavelength at which energy is being emitted it is therefore possible to distinguish the presence of all sorts of extremely rare components, which are in the sample only in the most minute proportions. These 'trace' elements are often metals or 'rare earths', metals like manganese or molybdenum, rare earths such as europium, scandium or samarium. The proportions in which these elements – as well as the commoner constituents – are present give a material a 'fingerprint' which can be compared with possible sources of origin, or with comparable artefacts, with great precision because the wavelength of each atom's radiation is very precise and because the amount of the element present is shown by the amount of radiation emitted at that particular wavelength. The method has the great advantage that it is 'non-destructive' and museums can have their exhibits back after they have been analysed.

One reason why archaeology has come rather late to the field of neutron activation analysis may well be that so many other people are desirous of using the comparatively rare and expensive facilities in nuclear reactors that make it possible. The technique is being increasingly used by crime detection forces, since it can analyse broken glass from a suspect's clothes or paint from the bumper (fender) of a car thought to have been involved in accident or hit-and-run crime. E. V. Sayre and his group at the Brookhaven

National Laboratories in the U.S.A. were the pioneers of using the technique for archaeological purposes from 1956 onwards and they have solved many problems, especially in the difficult fields of Central American pottery. For instance, neutron activation analysis settled the dispute about the origin of the Mayan pottery called Fine Orange Ware; it showed that the Kixpec site was the centre for the manufacture of this ware and that the samples found at Piedras Negras and elsewhere must have been imported from Kixpec, because the undoubted local pottery of Piedras Negras had quite a different composition in terms of trace elements.

These modern methods, however, have in no way contradicted all the results from earlier methods or even replaced them for many purposes. The first major archaeological study in which chemical analysis played a part has remained unshaken in its results; this was the study of British stone axes, which was co-ordinated by Professor F. W. Shotton. This study is still going on. Using the ordinary geological methods, called petrological examination, it has been possible to show that these tools, or weapons, of prehistoric Britain are made of at least twenty different types of stone, and many of the hand-axes dug up by archaeologists belong clearly to one or other of these groups of stone. So clear is the identification of the different types that in four cases it has been possible to find the original stone quarries from which the axes were manufactured. One of the best known is the site on the Langdale Pikes, a group of mountains in the English Lake District. Two others are in Caernarvonshire, north Wales, at Graig Lwyd and Mynnydd Rhiw (the most recently discovered). The fourth is in Northern Ireland at Tievebulliagh and Rathlin Island. In modern opinion the Langdale axes are the finest, and they are widely distributed throughout Britain, so perhaps our ancestors thought the same. The outcrop of rock from which the axe stone was quarried is still visible above the screes, two thousand feet above sea-level, and the working camps where the axes were shaped out of the rock pieces have been found below. Technically this Langdale rock is described as a 'fine-banded epidotized andesitic ash', which can be translated as meaning that it was formed from a shower of hot volcanic ash that fell into the sea of the Ordovician period.

In the case of some of the other groups of axes the parent stone can be found in outcrops in just one place in the British Isles, but the factories for turning the rock into axes have not been located. These sorts of studies have been followed elsewhere, notably in north-west Europe, and Palaeolithic axe factories have been found in other countries.

But having found the source and origin of a particular group of stone axes which have been picked up at many different sites, the geologist then hands back to the archaeologist, and the archaeologist can turn the evidence upside down so that his axes, now that he has found their origin, become clear markers of trade routes stretching out from the place of origin. He can then trace British axeheads onto the continent of Europe, and European axeheads can be found in Britain. The establishing of trade routes tells more than just the fact that trade went on and by what routes; it establishes levels of sophistication among the people trading, and leads to further useful questions such as what goods the buyers of axeheads had to offer the sellers.

The problem of the stirrup-jars of Thebes also shows what can be done with comparatively simple methods of chemical analysis. Stirrup-jars are large pottery jars which were used for the normal trade purposes of carrying and storing oil and wine throughout the Aegean and East Mediterranean in Bronze Age times. In 1921 excavations at Thebes in Greece revealed twenty-eight of these jars in a storeroom or passageway of a Mycenean palace under the centre of the modern city. On some of these jars were inscriptions in 'Linear B' writing, that is to say a script which was basically the one used by the Minoan civilization of Crete based on the enormous palace complex of Knossos, but which had been adapted for writing in the Greek language. (No one has yet deciphered the original 'Linear A' script.) Throughout the 1960s there has been furious controversy among scholars about the correctness of the date, 1400 B.C., given by the excavator of Knossos, Sir Arthur Evans, for the destruction of Knossos. This controversy stretches out to include the problems of the dates of the destruction of the Mycenean palaces at Mycenae, Pylos, Tyrins and Thebes, and also to the assertion by Sir Arthur Evans that writing died out after the destruction of Knossos. Evans's own excavations were originally started in an attempt to disprove the idea that Cretan civilization of that time was a mere dependant on the Mycenean civilization of mainland Greece, which had been revealed by Schliemann's dramatic discoveries of the Mycenean palaces and the 'treasure of Agamemnon'. Evans also wanted to prove that his Cretan civilization was literate. He succeeded in his original intentions and showed, in fact, that the Mycenean civilization of Greece was dependent, at least at first, on Crete. But then he wished to see, in the fall and destruction of Knossos, the virtual end of civilization in that part of the world. Work throughout the Aegean since Evans's time has shown that, while things started as he said, once Knossos had fallen, Crete took a very small part in a continuing civilization centred on mainland Greece. There are problems however, notably the remarkable likeness of the palace records of Pylos, on clay tablets written in Linear B, and those of Knossos. But since Pylos was destroyed only about 1200 B.C. some scholars have suggested that Knossos fell at that time, too – some two hundred years later than Evans proposed. The controversy still rages.

Some of the Linear B inscriptions on the Thebes stirrup-jars are linked with Linear B records found in the palace of Knossos, and so it has been widely assumed that the Theban jars came from Knossos as part of the vast trade system controlled by the huge palace in Crete. In view of the Knossos controversy Dr H. W. Catling re-examined the Theban jars in 1964, using the latest methods of spectroscopy to measure the proportions of nine elements in the pottery, including elements which the layman does not normally associate with pottery such as titanium, manganese and nickel. He then compared his results with pottery of fourteen different 'types' which came from different areas, or known pottery-producing sites, scattered throughout Greece, Crete and the shores of the Aegean. The composition of each of these types of pottery was likewise measured in terms of the nine elements and their proportional representation in the clay.

The results were clear-cut. At least seventeen of the jars, in two different groups, seemed to compare in their make-up only with pottery from east Crete, particularly from two known sites, where ancient cities once existed, Palaikastro and Zakro. A few of the jars seemed to come from Thebes itself; some others from the Peloponnese; and some did not match any known pottery. Two important features stand out further – none of the jars came from Knossos yet some of the east Cretan jars had Linear B writing on them.

From the point of view of the Knossos controversy these results please the extremists on neither side. Other excavations have shown Thebes to have been an extremely important and wealthy trading city. The jars suggest that Crete continued to trade quite satisfactorily and to have some influence after the destruction of Knossos. They also imply that writing in Linear B continued after the fall of Knossos and they definitely contradict the idea that in Crete Linear B writing was practised only in Knossos. The lessons of the Theban stirrup-jars would seem to be the common-sense ones, that great events like the destruction of Knossos do not end whole eras abruptly – the balance of political power will have changed, but some remains of the old power are still left behind, and trading and business carry on outside the immediate battlefield.

The importance of obsidian to Stone Age cultures because of the sharp cutting blades it provided has been mentioned earlier. The application of spectroscopic analysis to obsidian to find the sources from which it came and the routes along which it was traded has produced some of the most important archaeological results of the last five years.

Onion Portage, a site on the banks of the River Kobuk in Alaska, is potentially one of the most significant archaeological sites in the world. All the inhabitants of the Americas before the coming of the Europeans are believed to have their origins in peoples who crossed from Asia into Alaska at times when there was a land-bridge across the Bering Straits. But the rise in water levels that destroyed the land-bridge has, of course, destroyed the traces of human passage, while the frozen and thin-soiled country on either side of the former bridge offers few prospects of finding deep and ancient deposits. Onion Portage, which gets its name from the vast quantities of wild onions that grow there, is one of the very few important sites found in these Arctic regions and the depth of deposits may eventually take the record back 15,000 years. It has already provided significant links, in its early levels, with discoveries of stone tools similar to those made near Lake Baikal in Siberia and in Hokkaido Island in Japan. But the obsidian flakes found at Onion Portage remained a puzzle from the time the first excavations were made there by J. L. Giddings and Douglas Anderson of Brown University in 1961. Giddings thought the obsidian must come from the Aleutian Islands, and the varying amounts of obsidian in different levels at Onion Portage should then have represented the ebb and flow of trade relations between the coastal cultures, where people depended on fishing and hunting the arctic sea-mammals, and the inland cultures up the river valleys, where people lived by hunting caribou.

In 1970, however, two geologists, Patton and Millar, found the source of the Onion Portage obsidian at a place called Hughes, barely one hundred miles from the riverside site. The ratios of sodium to manganese in the obsidian confirmed the relationship which was evident on first observation. The artefacts at Onion Portage resembled the outcrops at Hughes in colour, opacity and lack of hydration on the surfaces; the hydration-dating technique showed that hydration is very slow in cold climates. The two geologists found a few artificially chipped flakes of obsidian at Hughes and some other chipped pieces on hilltops and ridges near by, places to which only humans could have moved them. Being geologists, working with other primary purposes in mind, they could not make a systematic search for prehistoric 'quarrying' or 'manufacturing' sites, but their geological studies showed that there may well be other sources of obsidian in the Yukon River basin, which would also be quite close to such coastal prehistoric sites as Cape Denbigh.

In a sense it is a negative finding, that there are local sources for obsidian finds in Alaskan sites, but it will undoubtedly help to clear up many problems in establishing the prehistory of one of the most significant of all migration-routes. The words of an editorial comment in *Nature* discussing the significance of the Patton-Millar discoveries are also worth recording:

The direct application of geological exploration to archaeological problems is still comparatively rare, however, probably because contact between students of the two disciplines is a matter of chance rather than a systematic attempt to discover and cooperate on common ground. How Patton and Millar came to see their geological investigations as a solution to a problem in Alaskan archaeology will perhaps never be known, but archaeologists should be pleased that the connexion was made.[3]

It was a team of two geologists, J. E. Dixon of Cambridge University, and J. R. Cann, of the British Museum, Natural History, and an archaeologist, Colin Renfrew of Sheffield University, which achieved similar results by similar methods in the Middle East and the Mediterranean. In the Old World, however, the interpretation of the results is very different because the archaeological record shows obsidian in so many completely different sites scattered over a much wider area. All the very early examples of settled communities mentioned in previous chapters – Jarmo, Jericho and Ali Kosh – and many other sites from Turkey down to Cyprus and Palestine and right across to Iran, produced at least some examples of obsidian tools. Places like Cetal Huyuk in Turkey, which was so well-developed by 6000 B.C. that it must be called a town, have provided sophisticated and beautiful obsidian daggers and polished mirrors.

This team, which began work in the early 1960s, started by studying a comparatively small number of samples of obsidian that had been excavated from sites in the western Mediterranean and Italy as well as in the Middle East. First they established that the ratio of the elements barium and zirconium was the most powerful measurement for discriminating between samples of obsidian from different sources in the Middle East. The precise amount of strontium was often a useful discriminant, also. Then they applied spectroscopic-analysis to a very large number of obsidian samples found in Middle Eastern excavations, and they found that they could place every single sample in one or other of eight groups, the groups being defined by the precise ratios of the amounts of 'trace' elements in them.

It was already known that obsidian could be found occurring naturally in two different areas of extinct volcanoes in central Turkey and Armenia. The British team therefore examined the chemical composition of freshly mined obsidian from these mountains, and found that the large majority of the samples of obsidian excavated by archaeologists from sites all over the Middle East corresponded exactly in chemical composition with obsidian from the mountains. A further search in the mountain ranges that extend from central Turkey to Armenia enabled them to find another source of 'natural' obsidian which corresponded to many more of their samples of excavated obsidian from archaeological sites. It is clear that there must be further sources of obsidian still to be located precisely, but the vast majority of finds of archaeologically significant obsidian in the Middle East can now be said to come with a fair degree of certainty from two general areas in the Turkish highlands. Many can be assigned definitely to particular quarries in these two areas.

Much more information came from this study. By comparing the places where the obsidian had been discovered as part of human dwelling sites with the places from which it had originated it was possible for Renfrew and his colleagues to draw up rough maps of early obsidian trade routes. The chemical analyses showed that obsidian found by archaeological excavation in Turkey, Cyprus and Palestine all came from the source area in central Turkey. But obsidian found in early Mesopotamian and Persian sites all came from the Armenian source area. Sites in Syria, however, provided an overlap area, with obsidian from both sources.

To establish the main lines of flow of the obsidian trade in the prehistoric Middle East is an achievement of major importance for prehistory in general. But there is even greater significance in this study. If a place like Jarmo, a village of no discernible historic importance in the early stages of its existence perhaps as long ago as 5000 or 7000 B.C., could contain obsidian from a source many hundreds of miles away, then the amount of contact and trading between the little villages and communities of the Fertile Crescent in those millennia when agriculture was first being developed must have been far greater than we had previously thought. Any theories about the origins of agriculture which speak of 'independent' settlements or villages, or 'independent' developments in different places within the Fertile Crescent must surely be modified in the face of the evidence of the obsidian, which shows that there must have been contacts, and probably regular contacts, between the different communities over ranges of hundreds of miles.

Even more recently these studies of obsidian by Renfrew and Dixon have been extended to provide evidence that man had become a sailor and navigator of the deep seas as early as 7000 B.C. This fascinating discovery comes as a 'spin-off' from the excavation of the Franchthi cave in southern Greece by Dr Thomas Jacobsen and a team from the University of Indiana.

Franchthi cave is a very large natural structure, nearly 500 feet long and 130 feet wide, on the northern headland of the Gulf of Argolis. The major importance of the excavation there is the light it sheds on the prehistory of Greece as a whole, for the archeologists have discovered successive layers of habitation of Palaeolithic (Old Stone Age), Mesolithic (Middle Stone Age) and Neolithic times. Such a sequence is unknown anywhere else in the Old World, and the evidence for inhabitation in the Mesolithic is particularly important. Up to now there has been an extraordinary gap in the Greek record with no previous discovery of any inhabitants of the period from 9000 B.C. to 6000 B.C.

In a level which has been radiocarbon dated to between 7500 B.C. and 7000 B.C. Dr Jacobsen's team came across the first specimens of obsidian in the cave. In the same layer they also found fish bones much larger than any found on lower, earlier, levels.

Renfrew (who is now at Southampton University) and Dixon (who is now at Edinburgh University) subjected the Franchthi obsidian to a whole series of analyses in their laboratories in Britain and concluded, unequivocally, that it must have come from Melos, a volcanic island in the Aegean at least seventy-five miles across the sea from Franchthi. The conclusion must be that the people who dwelt in Franchthi cave before 7000 B.C. could cross the water for supplies of obsidian, and also used the water for deep-sea fishing.

Incidentally Franchthi cave provides evidence of a sudden change of culture about 6500 B.C. with the sudden arrival of domesticated goats and sheep along with cereals such as wheat and barley, which had not appeared beforehand even in their wild forms. Pottery arrived at the same time, too. This, of course, provides a date for the Neolithic Revolution arriving in southern Greece.

There has been yet another development from Renfrew's study of Middle Eastern obsidian. In this latest work he has compared the amounts of obsidian recovered by archaeology from various sites in relation to the distance of these sites from the sources of obsidian. This has led him to what may be even more significant reinterpretations of the archaeological record. But that study is not directly concerned with chemical analysis, and it belongs in a different branch of the stream of intellectual development of archaeology. It will therefore be treated in Chapter 16.

Chemical analysis is a rapidly changing and developing technology. New devices and techniques are being continuously introduced into scientific laboratories. Many of the new techniques have been specifically developed to suit the needs of particular scientific disciplines; electron microprobe analysis, for instance, has been largely developed for the benefit of the metallurgist. X-ray fluorescence analysis can be found in the laboratories of both the medical research teams and the geologists. Even more recently has come 'laser microprobe analysis'. No longer are these developments in other sciences ignored by the archaeologist. All these techniques are being used, even if only experimentally, to see if they can extract further useful information from the evidence in the archaeological record. The standard techniques of analysis are being refined and sharpened, too, and applied to many different materials found in the process of excavation. Hair, leather, bone, parchment and metal artefacts have all been made to give up extra information by careful analysis.

There is no doubt that this is one of the fields in which the application of laboratory techniques to the problems of archaeology is making the greatest impact at the moment. It can be reasonably expected that this impact will increase in effect in the years immediately ahead.

1. L. Woolley, *Digging up the Past*, pp. 32–3.
2. J. and C. Hawkes, *Prehistoric Britain*, p. 163.
3. *Nature*, vol. 228, 10 October 1970, p. 112.

11

Computers and the Past

It was as recently as the summer of 1971 that the computer first actually went into the field with archaeologists. Previously the computer had remained in the background, being used only when the excavators returned to their universities for the detailed study of what they had discovered. But when the Arizona State University Summer Field School in Archeology began its 1971 dig at a pueblo site on the Navajo Indian Reservation south of Window Rock, Arizona, there was a computer terminal with them at their base at Hunters Point B.I.A. School. Using an ordinary telephone line the terminal was connected daily to the Honeywell computer at the Arizona State University Computer Center at Tempe, three hundred miles from the diggers' base.

The field work consisted of a survey of some twenty different sites in the area, including a rincon and various nearby mesas. One of these areas, containing a complex of at least twenty rooms, a kiva, and the associated rubbish dump (trash mound in American), was excavated in detail, and dated from A.D. 925 to 1325.

The object of taking the remote computer-terminal with them was to see whether or not computer facilities could be made available for such an operation so far from base, and out in the field. They wished to see how the techniques, worked out back at the base laboratory, would cope with a real-life, real-time situation of archaeologists sending in their data at the end of each day's work and demanding instant computer analysis of their results to guide them in their next operation on the site. Other major objectives of the experiment were to see whether the average archaeologist, completely untrained in computer programming, could feed into the computer information which the computer could understand and use. This involved the development of a special language which the archaeologists could be taught and which the computer could process. Finally, there was the biggest question of all: would the computer really help the archaeologist in making decisions about what to do next and in interpreting what he had already found?

This is to look at the operation from the archaeologists' point of view. At the computer end of the operation, there were three major blocks of programmes. The first was intended to deal with the problems of untutored users; programmes that checked the daily input from the archaeologists for consistency and for correct use of the 'archaeological computer language' that had been developed. These programmes immediately sent back to the remote terminal edited versions of the data that had been fed to the computer, written in such a way that the archaeologists could quickly check for good sense. The second major block of programmes sorted out every digger's data and merged it all and stored it in a huge, but easily available 'file'. The main value of the programmes to the archaeologists lay, however, in the third block of programmes, though to the computer-specialists these were probably the least interesting, because the most routine. This last block of programmes allowed the men in the field to do two main things – they could ask the computer to get out of the main 'file' a list of all the items of any particular type and perform statistical tests on these items, and secondly they could demand logical searches of the main file. In other words the men three hundred miles from the computer could ask each day about the significance of what they had found in terms of all the rest of the work on the site up to that moment. On this basis they ought to be able to make better decisions on what to do next: whether to continue to work in the same way or whether to change their objectives; and the logical searching would rapidly show whether a preliminary hypothesis was being verified by the results of the dig or not.

Incidentally, the archaeological computer language in which all this 'conversation' went on used words with which the archaeologists were familiar, and had a syntax which combined the computer's need for strict logic with a simplicity which enabled the man in the field to learn it quickly. Above all the archaeologists did not have to learn any special form or set of symbols for coding their results to feed into the computer.

The first results of the experiment showed, not surprisingly, that a small dig was not the ideal way of making use of the power of the computer; there was not really the volume of data nor a true requirement for speed in response. But it was equally shown that the verification tests provided by the statistical analyses of data, which the computer could perform so quickly, were definitely helpful in guiding the men in the field, confirming or rejecting or modifying their hypotheses with great ease. The logical search facility had the peculiar, valuable, but sometimes disconcerting habit of showing up with great rapidity areas where there were gaps in the evidence that had been collected; and thus suggesting, by implication, where further digging should be undertaken. The main conclusion, then, was that such computer facilities could be very valuable, particularly in cases of large-scale rescue digs, where there was a great deal of data and a real need for speed, not simply in dealing with the data, but more in dealing with the provisional hypotheses of the men on the job, and directing them where to go next. Not least, the experiment showed that the cost of such an aid to archaeology is not unreasonable.

This use of a computer by archaeologists actually in the field is, however, an entirely new departure. For the most part computers have been used in archaeology mainly for a reassessment of data already gathered. It would be more accurate, in fact, to write of the use of mathematics in archaeology and to distinguish that from the use of computers. The Arizona field trial is a clear case of the use of a computer in its own right as an aid to archaeology; it has deliberately attempted to bring the inherent power of computers in filing, sorting, dealing with large quantities of data, and providing rapid answers, to the aid of the archaeologist. This can be compared with the application of mathematical techniques on archaeological data, even where those techniques require a computer to operate them, because no man could face the enormous amount of time and drudgery involved in performing the calculations by hand. There is no attempt here to imply that this second category of computer-use is less important to archaeology; it may be less interesting for the computer programmer, but till now it has certainly proved of more value to the archaeologist.

There are, indeed, some very good examples of cases where the use of mathematical techniques which require a computer for their application have wrung completely new information from collections of archaeological data, even when the data have been perfectly well known and widely studied by traditional methods for many decades. The Czech archaeologist Evzen Neustupny,[1] the man who fought for the 'long chronology' for the European Neolithic in the debates over the recalibration of radiocarbon dating (see Chapter 4), has recently turned to working over old collections of material with new techniques, and has called in the programmers of the Centre of Numerical Mathematics at Charles University in Prague to help him. His reasoning was based on a conviction that 'in some cases archaeological knowledge has not notably increased in recent decades despite the rapidly growing volume of archaeological evidence available'.

Neustupny took a well-known feature of the archaeological record of central Europe as his working ground; this was the Corded Ware culture of Bohemia. This culture is known only from graves, but it is clearly the culture of a people who invaded a wide area of central Europe from Russia and went on to reach Denmark and the shores of the North Sea. These people are widely known, also, as the Battle-axe people because of the superb polished stone axes, with holes in them to take wooden hafts, which are the great glory of their graves. They conquered the 'first farmers' of northern Europe when they swept in around 3000 B.C. To the archaeologist, however, they have traditionally been known as the Corded Ware culture from their habit of decorating their pottery with patterns formed by pressing ropes into the wet surface of the clay before it was fired. Nearly a thousand of their graves, usually burials of a single person under a fairly simple mound or barrow, are known from many parts of northern and central Europe, and three hundred sites have been recorded in the Czechoslovakian province of Bohemia. In particular there is is a huge Corded Ware cemetery containing more than 120 graves of Corded Ware people at a place called Vikletice in north-western Bohemia.

Now the Battle-axe people, not unnaturally, buried the battle-axes with dead men, whereas in the graves of women they put no weapons, but only ornaments and pottery. They put pottery in the graves of men too, but more important they buried men lying on their right sides and women lying on their left sides. This we know, from at first guessing that graves in which battle-axes were found would be warriors' graves, and then from noticing the different positions of the skeletons. But the great expert on the Bohemian Corded Ware culture, Dr M. Buchwaldek could say little more than this even after a quarter of a century's detailed study of the graves and their pottery, ornaments, axes and skeletons. In the few cases where they were able to give definite answers, the anatomists had confirmed that the skeletons lying on their right sides, with axes in their graves, were indeed male, and the others, lying on their left sides were female. Dr Buchwaldek was able to add that a certain type of small pottery cup seemed to be found in women's graves only along with some female ornaments. At the start of his work he produced a scheme for putting the graves in a chronological order according to certain differences in the pottery found in them, but later he abandoned this in favour of explaining the pottery differences in terms of there being two different racial groups among the Battle-axe people. This lack of real progress in discovering anything about the Battle-axe people despite the plentiful remains they left behind is the gist of Nuestupny's complaint against the inadequacy of the traditional methods of studying the archaeological record.

So he approached the problem afresh, with the promise of computer aid. He took two samples of graves – 120 from the large cemetery at Vikletice and 138 from other parts of the country, all of which had been well excavated. He divided the material from these graves into ten types in which he was interested – such types as beakers, jugs with handles, small cups without handles, cups with one handle, cups with two handles, the famous battle-axes, stone tools which could have been hammers, axes or maces, personal ornaments, and so on. All this material was available as 'published' material, which is to say it had been fully counted, measured and described by the archaeologists who had dug it up. They had published their findings in the recognized academic journals and books. In principle Neustupny did not have to touch or look at the pottery and stones themselves.

Taking the two sets of graves as two different samples from the same population, and taking the battle-axes as signs of male graves and the personal ornaments as signs of female graves, the other material was then studied and fed into the computer. Three sorts of statistical analysis were used, multivariate factor analysis, principal components analysis, and cluster analysis, which are all fairly standard computer techniques.

Broadly speaking, these analyses find out whether any particular type of object is concentrated in association with any other type of object, or whether they are scattered at random throughout the whole collection. In other words, the computer will find groupings of similar objects, turning up in association with each other time and again, where the human being could not spot such groupings because they occur among so many objects all told. This power of the computer to reveal patterns of recurrence can then be interpreted by the archaeologist in other terms, perhaps in terms of the social relationships of the human beings involved, on the grounds that repetitive patterns of social behaviour will leave repetitive patterns of objects. Yet since human beings are not automata these patterns may be difficult to spot in the total assemblage of all the objects in the archaeological record.

In the case of the Battle-axe people and their Corded Ware pottery in Bohemia, the computer showed up some extremely interesting patterns among the relics. First, it showed a fairly clear time-pattern – a steady change in the types of pottery used, which could only reflect the changes of taste and manufacture through time. So now the archaeologist knows which types of pottery came first, and which at later stages of the culture. But even more striking differences emerged between the two sexes. Males were not only buried with their battle-axes but with whole sets of pots, jugs, vases and cups which seem to have been used exclusively by males. Similarly certain types of pots and implements seem to have been exclusively for female use; thus amphorae with a certain type of engraved pattern are only found buried with females, while amphorae in general with all other types of decoration seem to have gone into the graves of both males and females.

The conclusion about this people, who are so important in the origins of European races and cultures, is that there was strong sexual dimorphism; in Neustupny's words, 'There were two distinct subcultures in the Corded Ware culture, one male and the other female.' What seems even more unusual, this distinction between the male culture and the female culture extended even to young children. Whereas many tribes and societies bury their children with quite different ceremonies and rites from the adults, among the Battle-axe people, what mattered was whether you were male or female, not how old you were. Neustupny's researches showed children as young as twelve buried exactly as adults, male or female being the distinguishing feature.

The essence of Neustupny's work was the use of the mathematical power of the computer, and by applying that power to data collected by previous archaeologists, he revealed details of the social organization of groups of people living five thousand years ago. Similar uses of computer power working on data collected by other diggers have recently been reported on such different subjects as palaeolithic stone tools in Italy and southern Europe. On the opposite side of the world, computers have been fed previously collected data in an attempt to sort out the development in time and space of the various types of stone axes found in New Zealand and the South Pacific islands.

If, on the other hand, it has been decided to use computer techniques before an excavation begins, then obviously the excavating methods will have to be devised to fit the computer programming. This may involve a clear decision to excavate in one way rather than another.

In their attempts to cast light on the social order and daily life of prehistoric men, archaeologists have recently been giving a great deal of attention to the actual position in a site in which they find man's artefacts. The idea – which springs largely from the work of François Bourdés, the great French expert on stone tools – is to try and reconstruct the 'tool-kits' of early man. It is often impossible to tell just what a stone tool was intended for, what job it was designed to do. If, however, a certain type of flint implement is found only in the immediate area of the charred earth and stones that indicate a hearth – and it is found nowhere else in that particular cave or site – then it is reasonable to deduce that that particular type of implement was used in cooking and the preparation of food. It may also turn out, at that site or some other, that this particular type of implement is associated, therefore, with women's work, and hence with women in general. Other implements, or more accurately groups of implements, can reasonably be associated with the work of dealing with skins and hides of slaughtered animals, and hence with the making of clothes, ropes and so on: bone needles, borers and scraping tools would be typical of such a group. With this type of technique it has been possible to establish that certain sites were used as 'kill-sites', essentially hunting camps, occupied only for short periods of the year and not constituting the main base, or 'living-site', of the men involved.

There are two widely used mathematical techniques by which archaeologists establish whether certain types of tools are congregated in particular areas of a site or whether they are merely scattered about at random. (Most of the results obtained so far have in fact been based merely on observation and the general impression gained by the excavators, but in this chapter we are specifically considering the application of objective mathematical methods.) These two methods may be called 'dimensional analysis of variance' and 'nearest neighbour statistics'. In reality the second involves plotting the co-ordinates of every find, measured from some point of origin, and then applying standard statistical techniques to find out where there are clusters of like objects. The first of the two methods involves dividing the site up into imaginary squares on a grid and plotting the number of objects and their frequency in each square; originally this technique was invented by botanists to get computer analysis of the difference in the frequency and population of plants over an area in which they were interested.

This spatial analysis technique of grid plotting was the one chosen for the excavation of a very small cave at Guila Naquitz in the Oaxaca valley of Mexico. Kent Flannery, whom we have already seen at Deh Luran, directed the excavation. At the very lowest levels of occupation his team came across the traces of small, seasonally used camps of a group of men who were hunter-gatherers – that is to say they had no agriculture and they preceded the Neolithic Revolution of Central America. The wider importance of the Oaxaca valley is its strategic position in Central America – more precisely in what the archaeologists call Meso-America – between the high plateau valleys of central Mexico, where the cultures and cities of Teotihuacan and the Aztecs flourished, and the jungle lowlands of the great Maya civilization. The Oaxaca valley contains the ruins of Monte Alban, one of the first of the great Central American temple-cities to be discovered; it is believed by some to be the home of the Olmec culture, the first of the high cultures of the New World. Among the archaeological finds at Monte Alban there are items showing the influence of both Mayas and the inhabitants of Teotihuacan, yet the city itself seems to be part of the less well-known Mixteca 'empire'. Despite this obvious importance, as recently as 1969 Friedrich Katz, Professor of Latin American History at the University of Chicago, wrote:

Archaeological research, in spite of the great progress that has been made recently, has not advanced so far in the valley of Oaxaca as in

Teotihuacan. One of the consequences of this is that the date of the beginning of human settlement in Oaxaca is still in doubt. Because of Oaxaca's proximity to the valley of Tehuacan where MacNeish discovered traces of human settlement going back to 7000 B.C., the conclusion was drawn that the valley of Oaxaca must have supported human settlement for approximately the same time. Nevertheless, incontestable evidence of very early human settlement has so far not been discovered. As in other parts of MesoAmerica the first signs of a high culture in the valley of Oaxaca likewise have no recognisable origins.[2]

Flannery, therefore, had several major objectives in sight when he dug at Guila Naquitz, but the particular application of the computer to spatial analysis of the lower levels was aimed to throw light simply on the behaviour pattern of the earliest inhabitants. Robert Whallon Jr recently reported on what they found on just one of these occupation levels and how they interpreted their results.

They found eighteen separate classes of remains in significant quantities. One class, naturally enough, was flint tools and debris; the rest were all plant or animal remains from the hunter-gatherer activities of the inhabitants. The spatial analysis of these remains was quite clear; nearly every single class was found to be grouped together – that is to say if the computer looked at any particular item 'x' (perhaps remains of the prickly pear) then all the examples of 'x' tended to be concentrated in one particular part of the floor of the small cave. Even more important to the final interpretation were the concentrations of items which were near the concentration of 'x'.

So in one part of the cave floor they found that all the remains of acacia and mesquite (technically both tree legumes), all the remains of the leaves (or stem section) of the prickly pear, and the remains of the agave, the century plant or maguey, were grouped together. The significance of this collection is only clear to someone who knows the botany and ecology of the Oaxaca valley. But with that knowledge it is clear that all these plants come from one zone of vegetation, the Lower Thorn Forest. In other words all these plants grow in the same kind of habitat in the foothills of the mountains. The obvious conclusion is that the hunter-gatherers went on trips to this particular part of the countryside and gathered together all the fruits and plants that they knew were useful to them and which grew in that particular part of their territory. But even further results were obtainable from this concentration of remains, for among them were many specimens of maguey which appeared to have been chewed and spat out. These were in fact 'quids', like quids of tobacco or betel-nut such as are chewed and spat out in our twentieth-century world. Apparently the early inhabitants of the Oaxaca valley chewed maguey while they worked together preparing and processing the vegetable food they had gathered on one of their foraging trips. Also found with this particular concentration of remains are the relics of a certain wild variety of cucurbita, a wild squash. It had been suspected that this particular ancestor of one of the first plants that man domesticated had been gathered in the Lower Thorn Forest, but never actually proved. The evidence from this cave at Guila Naquitz provides strong confirmation, if not final proof, that this plant came originally from the same habitat as all the others associated with it.

Another group of plant remains found together in the cave are susi nuts (jatorpha) together with the pods of two other tree legumes, both called *guajes* locally, but classified officially as *lucaneana* and *lysiloma*. All these plants also come from the Lower Thorn Forest habitat, but are usually found rather higher up the slopes of the foothills than the first group. Presumably since they were found grouped in a different area, special collecting expeditions to the higher slopes were organized to get these plants and fruits.

Another quite clear grouping on the cave floor, as seen by the computer, corresponds exactly to the plants of a different vegetation zone, the Upper Oak Woodland. Acorns, pinyon pine nuts, hackberries and the wild bean that grows in the underbush of this type of forest, were all found together. On the other hand, another inhabitant of the Upper Oak Woodland, the West Indian cherry, or *nanche*, was found in a different part of the cave. This was presumably because the *nanche* fruits come earlier in the season than most of the other plants in this zone. So again we have evidence that the men went on special trips to special areas for the wild harvest they depended upon, and then dealt with the results of that particular trip in some special way when they returned to their base.

The meat they ate came from turtles, found in the ponds on the valley floor, from deer, found in the Upper Oak Woodlands, and from rabbit, found everywhere. But though these creatures are found in different habitats they were butchered and processed together in the same area of the cave. This difference of treatment from the plant harvest can be interpreted in terms of hunting being the occupation of the males, while plant gathering was done either by the women alone, or more likely by groups of both sexes.

The remains of flint tools were found with some of the concentrations of food remains and not with others. They were found among the animal remains and among the remains of plants from the Upper Oak Woodlands and the higher slopes of the Thorn Forest. But they were hardly found at all with the first group mentioned, the group from the bottom levels of the Lower Thorn Forest. In other words, tools were used for butchery and meat preparation and in dealing with some nuts and pods, but, quite clearly, not for all the foods these men gathered and ate.

Finally, there was one set of remains very much in a class of its own. These were the remains of the fruit of the prickly pear. Whereas the leaf and stem parts were found with the first group, the remains of the fruit were by themselves, associated with no other plant remains nor with flint tools. The fruit of the prickly pear (the cactus opuntia) was clearly of great economic importance for these early groups of hunter-gatherers, but why did they treat it differently from other foods by concentrating it in one particular part of their cave? A possible explanation suggested by Flannery is that it was so important that special gathering trips were organized, perhaps lasting several days, so that the largest possible quantities could be obtained.

The study of one occupation level of a single cave in Mexico is, obviously, only one small example of the use of the computer in archaeology. Its significance is that it shows the tendency for the archaeologist to change his method of work to suit the extremely powerful assistance the computer can give. Computers are being used to help solve problems as various as calculating the industrial production of early British pottery kilns, and encoding and classifying the entire corpus of sculptured inscriptions left by the Roman Empire. But the importance of computers lies not so much in the results they can provide, as in the changes in thinking and in modes of action which they force upon us.

One of the changes in methods of thinking which computers have brought to the field of archaeology has been to raise the question 'Did our ancestors have computers of their own?' This is a dangerous question because it asks us to play a time-wasting game, the numbers game. And the numbers game has been one of the greatest delusions on the fringe of archaeology ever since men first started wondering whether the precise dimensions of the pyramids of Egypt were a form of information-code in which were hidden the secrets of the past or even the mysteries of the future.

It is admitted by everyone that the Babylonians on one side of the world and the Mayas on the other had compiled extensive astronomical observations of the moon and the sun; that they could predict eclipses and that they did this on the basis of complex and accurate calendrical systems. To go further than this, further than the written or sculptured record, and to enquire whether the buildings or constructions of the ancients had astronomical significance in themselves, has been regarded by most archaeologists, until very recently, as a 'crackpot' study. But on the other hand it is reasonable and responsible to ask, 'How much science, and particularly how much astronomy, was known to the cultures of the past?' Such inquiry was rendered even more respectable when, in December 1972, an international scientific meeting was held on this subject jointly by the Royal Society and the British Academy in London. To this meeting came undoubted scholars and serious students of the question from America, Mexico, Australia and Great Britain. One of the speakers was the man who, more than anyone else, has raised this subject from the crackpot to the serious, Professor Gerald S. Hawkins, formerly Professor of Astronomy at Boston University and now at the Smithsonian Astrophysical Observatory in Massachusetts.

Hawkins really founded the subject of 'archaeo-astronomy' in 1963 and 1964 when he published articles in *Nature* – the most respected of British scientific journals – claiming that Stonehenge was, in principle, a vast computer for observing and predicting the points at which sun and moon rose and set at such critical times as midsummer, midwinter and the equinoxes. What captured public imagination about his work was that he had used a computer of the electronic type to prove that Stonehenge was a computer of another type. It had of course been known for centuries, ever since Stonehenge was discovered as an 'antiquity' in Tudor times, that the midsummer sun was seen to rise exactly over the isolated 'Heel stone' if the viewer stood in the centre of the Stonehenge circle. This axis from centre to Heel stone is also the axis of the main entrance to Stonehenge as marked by the prehistoric Avenue.

Much first-rate archaeological work has been done on Stonehenge, notably by Stuart Piggott and R. J. C. Atkinson, which has established the broad outlines of the history and construction of the place. There were three main stages of construction between 1900 B.C. and 1500 B.C., in the course of which the whole appearance of the structure was changed time and time again as the stones were moved from one formation to another, or as fresh sets of gigantic stone blocks were brought in.

This archaeological work showed that there were many more positions at Stonehenge than those occupied by the stones the twentieth-century visitor can see. There are positions where stones once stood, there are a few mounds, manmade but of unknown purpose, there are some holes that have never, apparently, contained stones. All told there are 165 positions of possible significance at Stonehenge, and these would yield 27,060 possible alignments, that is, lines of sighting from one position to another that might point to some such spot as the place where the midsummer sun rose above the horizon. Hawkins fed into the computer the co-ordinates of the 165 positions and asked the machine to calculate the 'declinations' given by the 120 most obvious alignments. (By regarding the sky as a hollow sphere covering the earth, and by drawing circles upon it corresponding to the circles of latitude on the earth, astronomers tell the viewer how much north or south he must look. This direction is called a declination and is measured in degrees north or south.)

A preliminary study of the computer's answers convinced Hawkins that none of these alignments had anything to do with the positions of the stars or planets, but it did seem possible that some of them might show the positions of the rising and setting of the sun and moon. So he got the computer to calculate the declinations of the rising and setting of the sun and moon in 1500 B.C. Viewed from Britain the sun has a maximum declination to the north in summer and reaches its extreme southerly declination in winter. The moon has a more complicated apparent motion, with two different maximal northern and southern declinations, swinging from one to another over a cycle of 18.61 years.

The result of these calculations was to show that twelve of the significant alignments of various Stonehenge positions pointed to one of the maximum positions of the sun and another twelve pointed to maximum positions of the moon. The accuracies of these alignments were not very high by modern standards: they were within a degree for the sun positions and within a degree and a half for the moon positions. But Stonehenge now is not in the pristine condition of 3,500 years ago; many of the stones have fallen or tilted, some have been re-erected, and several of the positions have had to be estimated. But a mathematical/statistical calculation shows that the odds are ten million to one against it being purely chance that the sun should be seen to rise, or the moon to set, exactly in the gaps between the enormous blocks of the Stonehenge 'trilithons'.

It was in 1961 that Hawkins first got his computer to do the calculations. He went back to the machine in 1964, by which time the response to his first publication had made him aware of three 'extra' Stonehenge positions; in fact the positions of former holes apparently once occupied by stones. These immediately showed alignments with the sun and moon positions at the equinoxes. Further consultation with atronomical tables (specifically van den Bergh's *Eclipses in the Second Millennium B.C.*) showed that an eclipse always took place when the winter moon rose over the Heel stone at Stonehenge. In addition it is known – and has been known since antiquity – that eclipses of the moon can only occur when sun and moon are opposite each other. So, although an eclipse, predicted by the rising of the moon over the Heel stone, might not be visible to viewers at Stonehenge, such a moon rise would warn the priests that the event was likely in the coming year. But since the moon's swing between its maximum declinations is in a cycle of 18.61 years, the winter moon would rise over the Heel stone apparently once every 19 years – until suddenly there would be an interval of only 18 years. Hawkins surmised that the priests or observers at Stonehenge spotted that a 56-year cycle – that is 19 years + 19 years + 18 years – gave them a steady time-cycle on which to predict moon behaviour. Then he realized that the most completely unexplained feature of Stonehenge was the circle of so-called 'Aubrey Holes', 56 in number. These 56 holes appear to the archaeologist to have been dug for no apparent purpose; they show no signs of ever having held wooden posts or large stones. Some contain the ashes of cremations, a number have a sliver of blue stone in the bottom, but there is no comprehensible pattern and they were undoubtedly filled in shortly after they were dug. Hawkins suggests they were a primitive computer, enabling the priests of Stonehenge to predict eclipses, and whether he is right or wrong he is almost certainly correct in claiming that his is the only rational answer to the problem of the Aubrey Holes that has yet been proposed.

It is, of course, possible to argue against Hawkins's interpretation of Stonehenge on what might be termed physical grounds – to say that with fairly large horseshoe-shaped structures of uprights and arches it is inevitable that several lines or alignments are bound to point somewhere near the several declinations of setting suns and rising moons and so on. In other words the argument is that if you draw enough lines some of them are bound to point to somewhere that is significant. Hawkins, however, at the London conference mentioned above, claimed that new photogrammetric air surveys of Stonehenge had shown that the alignments were more accurate than those in his original work. Looking in other parts of the world he found that at the enormous temple to Amon-Ra, the Egyptian Sun God, built beside the Nile by Thuthmosis III, the alignment of the main axis of the temple was exactly that of the rising of the midwinter sun, whereas most of the other Egyptian monuments are set perpendicular to the Nile. Presumably the worshippers of the Sun God were thus able to celebrate the return of the deity after his long decline.

But, on the other hand, aerial surveys of the mysterious lines drawn in the Peruvian desert at Nasca seem to show that the lines have no astronomical significance whatever. Professor Hawkins claimed to have struggled for three years to find some significance in these lines, but could not find any correlation with astronomical events. The lines themselves can only really be seen from the air, and then it is clear that they are more truly straight than the Pan-American Highway which cuts across the same stretch of desert. The lines have been dated by the presence of Nasca pottery (of periods Nasca II and Nasca III) to just before A.D. 100. The fact that their purpose remains a mystery should reassure archaeologists, according to Hawkins, that looking thoroughly at a physical phenomenon does not always produce significant results; therefore when significant lines are found in some monument there is less likelihood of its being pure chance.

The most distinguished and controversial practitioner of archaeo-astronomy in Britain is Alexander Thom, formerly a professor of engineering at Oxford University. For the past twenty years he has studied the stone circles, the menhirs, cromlechs and 'standing stones' which are to be found dotted round the western extremities of Europe. These monuments or constructions are clearly to be distinguished from the Megalithic graves which are the chief archaeological feature of the same general area. The menhirs and standing stones do not mark burials, though most archaeologists have felt that they probably represent an earlier stage of the cultures that buried their dead in the massive stone-lined tombs called 'Megalithic'.

As befits an engineer, Thom does most of his work with a theodolite. Some of the sites he investigates are so rugged and steep that he cannot always get a full-sized instrument to them, however, and then he has to compromise with a lighter and more portable version of the device. The essence of his work is that he has surveyed, and measured with great accuracy, scores of sites where there are collections of standing stones or menhirs, and he claims that this has shown that most of these sites are lunar observatories in which the stones show the 'line of sight' to some distant, but clearly defined, geographical feature on the horizon which marks the rising or the setting of the moon at the solstice or at one of its maximum or minimum declinations. Broadly speaking, the geographical feature – the sharp edge of a distinctively shaped mountain, a nick or cleft in the profile of a distant range of hills, or some similar landmark – is the 'foresight'; the main group of stones outlines the viewing point and there may be one or more 'backsights' marked by standing stones, which give the alignment to various foresights which show critical moments in the moon's movements through the sky. Thom also finds a smaller number of 'solar observatories', sometimes on the same sites as the lunar ones, which seem primarily to have fixed the days of solstice and therefore to have kept the calendars of these Megalithic astronomers.

The most impressive, and most famous, of the Megalithic monuments of north-western Europe are the great rows of standing stones at Carnac in Brittany, where there are 2,750 menhirs arranged in rows stretching for miles across the gorse-covered, rather barren and gently undulating countryside. Thom's latest work has been to take a team of workers under the command of his son to survey the monument, for which no explanation commanding general assent has yet been found. Thom's belief is that Carnac is the same sort of construction as the array of less obvious standing stones, arranged in rows in a fan-shape, which he has surveyed at the 'Hill o' Many Stanes' at Mid Clyth in Caithness in the far north of Scotland. He claims that both are vast geometrical constructions, a sort of Megalithic graph-paper, by which the moon-observers of four thousand years ago could extrapolate the true maximum declinations of the moon from the observed points of setting and rising.

He goes further even than this in his claims for Megalithic Man. He claims that there was a 'Megalithic yard', a standard measurement corresponding to 2.72 feet. Thom believes there was also a smaller unit, one-fortieth of the yard, a unit of 0.816 inches. This smaller unit was used in inscribing geometrical figures on the standing stones and other rocks – these are the figures known to archaeologists as 'cup-and-ring marks', and they undoubtedly exist in large numbers and are otherwise unexplained. Similarly Thom claims that Megalithic men knew a great deal of practical geometry such as the right-angled triangles, which we call Pythagorean, based on unit lengths of 3, 4 and 5, and 5, 12 and 13, (and possibly others as well). He also believes they knew the practical geometry of shallow arcs, and he believes they could draw both circles and ellipses, since he has found sets of standing stones which his surveys show to be very accurate circles and ellipses – drawings on the ground with the stones placed to an accuracy of 1 in 1,000. All this implies a background of technology of a sort we do not normally associate with man in his Neolithic or pre-Neolithic phase; a technology of levers, foundation building, and ropes, of mathematical thinking of some sort, and the use of accurate measuring rods and beam compasses with points set to within a few thousandths of an inch. To archaeologists brought up to think of technology beginning with the invention of pottery and weaving, this is difficult to accept.

Thom's opinion that there were Megalithic units of length has come under serious scrutiny; the difficulties of proving the existence of a quantum of measurement in Megalithic layouts by searching for such a unit among Thom's vast collection of measured distances between stones, was aired at the London conference in December 1972. But at the same conference there came archaeological confirmation of one of Thom's proposals.

Among the wild and beautiful hills and islands of south Argyllshire, Thom has found at least a dozen sites which he considers to be solar or lunar or 'solstitial/calendar' observatories. One of the most dramatic is at a place called Kintraw, on the west coast of the Mull of Kintyre, where a group of stones, a menhir and the remains of a cairn mark a place where the setting sun at midwinter disappears behind the mountain of Beinn Shiantaidh on the island of Jura, nearly thirty miles away across the sea. But on the right-hand side of this mountain, as seen from Kintraw, there is a steep drop in the skyline to the col between Beinn Shiantaidh and the neighbouring Beinn a' Chaolais. For a brief moment the sun is seen again almost as a flash of bright light in the bottom of the nick of the col before it disappears finally behind the second mountain. This gives extremely accurate alignments, or a very accurate calendar day for the midwinter solstice. But there is a problem, or rather there was a problem for Megalithic Man. The line of sight from the little plateau at Kintraw to the distant horizon is cut off by a low ridge. Standing on a small cairn (and there are ruins of a cairn) would get over the difficulty. But how to place the cairn in the right spot? For immediately behind the plateau with the stones is a stream in a deep gorge, a gorge which is virtually impassable without serious rockclimbing. Thom found what he thought had been Megalithic Man's solution: on the far side of the stream and only a few feet higher than the site itself was a little platform in the mountainside directly in line with the cairn, the standing stones, and the distant col in the mountains of Jura. From this vantage point the first rough observations might have been made and the building of the cairn in the correct spot could have been directed. (Though to reach this little platform involved a long walk downstream to the nearest bridge and a climb back above the edge of the gorge on the side of the mountain furthest from the true viewing site.) If Thom was right this little platform on the hillside was probably manmade, and if a traditional archaeological dig showed that it was manmade then Thom's interpretation of the whole site as an observatory lined up on the distant col was probably right, too. Ewan MacKie, of the Hunterian Museum at Glasgow University, decided to test Thom's theories in this way and he discovered that, though the platform showed no trace of human occupation or artefacts, it had probably been made by man. He reached this conclusion by analysing the precise orientation of the long axis of all the pebbles and stones which made the 'floor' of this platform; the pebbles were disposed in completely random directions, whereas if water, ice, or other natural forces of erosion had carved the platform from the hillside or deposited the pebbles there, one would have expected to find some prevailing direction in the orientation of the stones.

MacKie reported this finding at the London conference on 'The Place of Astronomy in the Ancient World', and his summary of his paper includes some words which reveal both the importance and the difficulty of Thom's work.

> An astronomical interpretation of the British standing stone sites has been developed in great detail by A. Thom, *using methods and data which have hitherto been rarely, if ever, used by archaeologists* [author's italics]. The unfamiliarity of these methods and the revolutionary nature of the conclusions drawn from them, have no doubt contributed to the difficulties which the profession is evidently encountering in coming to terms with Thom's ideas. However the raw data on which the theories are based are, like all other archaeological data, susceptible to checking and testing in traditional archaeological ways – by field-work and excavation.[3]

Indirect support for Thom's thesis came at the same London meeting from Professor R. J. Atkinson, of Cardiff University, who is not only one of the excavators of Stonehenge, but is also the greatest authority on Silbury Hill, a huge manmade earth mound, the largest construction in north-western Europe before the Roman times. Excavations have shown that the mound is much more than a simple pile of earth; there are several concentric, circular walls built into the hill to keep its shape. Therefore the builders, who spent eight thousand man-years of work with deer-antler picks and shoulderblade shovels in completing the mound, must have had some knowledge of what we now call soil-mechanics. Indeed, Professor Atkinson argues (along with Thom) that the very building of the Megalithic monuments, whatever their purpose, shows in itself a surprising degree of engineering skill. Without knowledge of fairly sophisticated techniques the great stones just could not have been transported or raised.

Furthermore, according to Atkinson, who has himself applied modern mining techniques to burrow into the centre of Silbury Hill, mining technology was highly developed in western Europe even earlier than Megalithic times. As early as 4300 B.C. the flint mines went deep, and burrowed through as many as ten inferior seams of flint before the flint-knappers started extending galleries when they found the sort of material they wanted. Very recently evidence has come out of South Africa that the Border Cave in the Lebombo mountains of north Natal was the scene of mining operations 50,000 years ago, so there is nothing ludicrous in the suggestion that man in western Europe could mine deep some 7,000 years ago.

Yet the problem for the archaeologist in all this is that Megalithic Man left no paintings or sculptures, and certainly no system of writing or recording his knowledge. In this respect he differed greatly from the civilizations of Egypt, the Middle East, the Far East and Central America. Yet the dates at which Stonehenge was laid out as an observatory and astronomical computer must be between 1900 and 1500 B.C., say 1750 B.C. If Thom's theories are right the astronomical data seem to point to the construction of these northern observatories even earlier, at least 2000 B.C. Yet these dates are as early as the first cuneiform texts reporting Babylonian astronomy, such as the appearances and disappearances of Venus, and the Egyptians undoubtedly took their astronomy mostly from Babylon. The Maya astronomy can only be dated to the first centuries of the Christian era.

There are even more disturbing ideas floating about in this field. The American, Alexander Marshack, has interested himself in early counting, and he claims that many famous examples of Magdalenian art found in Europe and dated to 15,000 B.C. show clear traces of counting marks. He distinguishes as many as a dozen different, recurring symbols marked on the bone and antler carved tools, which have so far gone unnoticed by the archaeologists who have been more interested in the art-forms and the typologies they could deduce from the styles. Marshack claims too, that the earliest 'notations' began among the Cro-Magnon hunters of the Aurignacian period dated to 32,000 B.C., though at that time the markings were normally made on separate stones or special marking surfaces inside caves. South Africans excavating the Border Cave in Natal, mentioned above, have found notched bones, which they believe to indicate counting practices, dated to almost the same time, 35,000 years ago.

This book is not primarily concerned with whether any of these theories turn out to be right or wrong – South African counting 35,000 years ago or early Scottish moon-watching 4,000 years ago may, or may not, have happened. The important point is that men who are not professional archaeologists – Hawkins an astronomer, Thom a civil engineer and Marshack a science writer before he became a Research Fellow of the Peabody Museum at Harvard – have looked at archaeological remains with modern mathematical and computer theories in mind, and found significance factors in the data. These factors are strong enough at least to demand explanation, and are likely to prove true enough to enter into the archaeological corpus of knowledge, at least to some extent if not entirely.

There are a number of important implications in this. First, it means that we have to revise our view of our remote ancestors; there is a clear implication that they could think very much as we do. Their mathematical skills as individuals were, at least potentially, as great 35,000 years ago as ours are now. Putting it bluntly, they were a sight cleverer than we thought. For the archaeologist this implies that he must look for other things in addition to the signs which he has been trained to observe traditionally. He must not only observe the demands of the physical chemist in obtaining samples for carbon dating, he must not only take botanists and plant-geneticists into account, he must now consider mathematical possibilities and bear in mind the technological problems that were solved in providing the data that he excavates.

1. The work of Evzen Neustupny and Robert Whallon, Jr, was reported at the Research Seminar on Archaeology and related subjects – 'The Explanation of Culture Change: Models in Prehistory' – held at the University of Sheffield, December 1971.
2. F. Katz, *The Ancient American Civilizations*, p. 73.
3. This summary was given at the meeting on 'The Place of Astronomy in the Ancient World' organized jointly by the Royal Society and The British Academy, London, 7 and 8 December 1972. The Proceedings of this Seminar have been published while this book was in proof – Gerrard Duckworth, 1974.

12

The Remains of Man

The proper study of mankind is man – an old tag, and like all old tags and Delphic oracles capable of more than one construction. In the sense that all archaeology illustrates the history or prehistory of man and his cultures, archaeology is a study of man. But if we regard man as a biological entity, then his biological remains, as opposed to his cultural ones, have told us surprisingly little about him. This is very strange, for not only do we know more about man, the animal, than we know about other animals, but also the remains of men's bodies, in graves particularly, are among the commonest items of data in the archaeological record.

Consider a passage from the introductory chapter to the massive 'round-up' of the current interactions between laboratory sciences and archaeology, *Science in Archeology*, edited by Brothwell and Higgs. Don Brothwell opens his section on 'Human Biology of the Past, with these words,

Perhaps the oldest of all 'specialist' fields associated with archaeology is that concerned with the analysis of the remains of man himself. Oddly enough, however, it is a field which seems to cause more unhappiness to the excavator than interest, especially to those who are unfortunate enough (they think) to come across the remains of many bodies. In the Mediterranean area Egyptologists have shown themselves to be the most calm and collected in the face of large numbers of burials and as early as 1910 large and detailed monographs were appearing on East African material. Excavators in Britain have certainly not behaved so commendably, and many skeletons have been lost, badly stored or thrown away.[1]

Perhaps one of the reasons for this surprising situation can be found in another sentence by Brothwell in the same section:

'Measurement is principally of value to demonstrate variability rather than close affinity.' It is probably this, comparatively recent, discovery of the immense variability in human shape and size that has caused the blockage, the lack of progress.

In the earlier days of archaeology there was indeed, as Brothwell says, much anthropometric activity, much measuring of skulls to decide whether they were long or broad or how much brain capacity they had; words like brachycephalic were thrown around and were considered highly significant. Races of men seemed very important. There were proto-Mediterraneans in the Near East, and European cave sites contained Negroid and Eskimoid types. Zeuner for instance devoted a whole chapter of his *Dating the Past* to the 'Chronology of Early Man and his Cultures', but by 1958 he was compelled to admit that no firm conclusions could be reached: 'The chronological distribution of early man does not conform to some of the current theories on the evolution of man . . . though the chronological arrangement cannot be regarded as infallible, it will require fresh evidence and not merely arguments, views or inclinations to dislodge substantially the more important examples used.' The crucial question for early times is the exact relationship between the cultures that are found and the arrival of modern man, *Homo sapiens*. Did the cultures of the Upper Palaeolithic or Mesolithic, i.e. the stages immediately preceding the Neolithic Revolution, belong only to *Homo sapiens*, or was modern man merely a modifying influence affecting the evolution of the cultures directly from the preceding Mousterian? Zeuner can only say, 'New discoveries are necessary to decide which of these alternatives or any new or unexpected ones, are right.'

Four years later Carleton Coon produced his massive book *The Origin of Races*, in which he drew together all the information then available, and drew the conclusion that *Homo sapiens* had evolved in parallel in the five basic races of mankind – the Caucasoid, the Mongoloid and so on. He specifically rejected the idea of discussing what this meant in terms of blood or brains and wrote in his Introduction:

Despite claims to the contrary, the blood groups of fossil bones cannot be determined. Nor can dead men take intelligence tests. However, it is a fair inference that fossil men, now extinct, were less gifted than their descendants who have larger brains, that the subspecies which crossed the evolutionary threshold into the category of *Homo sapiens* the earliest have evolved the most, and that the obvious correlation between the length of time a subspecies has been in the *sapiens* state and the levels of civilisation attained by some of its populations may be related phenomena.

But he has to qualify this.

Yet every major race, however advanced in civilisation some of its component populations have become, also contains remnant bands of simple hunters and gatherers to remind us all whence we came. The monkey hunters of the forested slopes of Central India are as Caucasoid as Charles de Gaulle, and the Ghosts of the Yellow Leaves who haunt the hillsides of Upper Siam and Laos are as Mongoloid as the Mikado.[2]

More than 650 pages later Coon admits that his ideas are not generally accepted and that the whole of his structure for the two billion inhabitants of the earth is based on little more than three hundred bone-bearing sites.

Since that was written in 1962, interest in this sort of subject has much declined, perhaps because of the series of discoveries by the Leakeys in East Africa, which have convinced many people that there is so much still to find that it is too early for large-scale theorizing. But equally important is the scientific realization of the fact that measurement of human remains is so far telling us little more than that there is an immense amount of variability in the human body – and this accords well with the broader biological view, current at the moment, that it is precisely in its wide genetic range of adaptability that the human species has proved so successful in evolutionary terms.

Some of the recent expressions of this uncertainty can be found, at least implicitly, in the various chapters on 'Man' in *Science in Archeology*. L. H. Wells, the Professor of Anatomy at the University of Cape Town, says 'The myth of the "gorilloid" build and stance of the classic Neanderthal type must now be considered completely exploded . . .', and he concludes:

At the present time the evidence for considering the earliest hominines to have been of pygmy stature is only tenuous and positive evidence for any early hominine or even hominid having been of giant stature is non-existent. It remains possible, therefore, that the extreme variations in stature encountered in modern *Homo sapiens* may have been developed within this species only in comparatively recent times.

Even dealing with what would appear to be the comparatively simple matter of sex determination in skeletons much more recent than those of hominids and hominines, Santiago Genoves, of the Historical Institute, University of Mexico, concludes by pointing out that there are now two different modern methods which are most favoured, and of these advances he comments: 'Perhaps, by following these methods we shall in future possess true limits of accuracy for the sex determination of various parts of the skeleton, and for a range of human populations both prehistoric and recent. This is a desirable goal which is not so far distant. This requires deep knowledge of mathematical procedures and of statistical techniques.'

The same author emphasizes the uncertainties now admitted in measurement:

It must however be said at the outset that recent studies have shown once more that, although we can arrive at a fair degree of exactness in age estimates, the variability within and between races is considerable. Indeed, there remains much to be discovered in this field of study, which means that the estimation of age in either individuals or populations is still subject to a variable margin of error.[3]

The examination of human remains for what might be termed 'gross' characteristics is an implicit acknowledgement of the fact that living human beings recognize individuals, and differentiate between different racial groups by such factors as stature, facial appearance, sex determination and so on. More scientific differentiation can be made by the use of other characteristics such as blood-groups and antigens. Blood-groups, the well-known 'A',

'B' and 'O' groups and the less well-known 'Rhesus' and 'M-N' groups, seemed a hopeful prospect twenty years ago, when it was known that the frequencies of such blood-groups did differ among the living races of the world. Although the latest methods enable blood-groups to be discovered from some skeletal remains, it is by no means always possible, as Carleton Coon pointed out, to discover them in every case. The conditions of burial affect the availability of this evidence.

The chief problems of studying the remains of man have therefore been the failure of most conditions of burial to preserve those features of the human body which we normally use for differentiating between individuals and classifying different types of living individuals, and the rather brusque treatment which archaeologists of the past have often accorded to human remains. There is also a less obvious problem – the lack of personnel to deal with such remains if and when they are found. The American, Calvin Wells, believes that a trained doctor or medical man is the best person to examine human remains, but there are others who feel that almost anyone trained in scientific techniques of measurement can deal fairly adequately with skeletons, at least when they are found in large numbers. From one point of view it is desirable to find large numbers of skeletons in the same burial ground, for then large numbers of measurements can be made and statistically significant statements can be produced about the stature, states of health and demographic structure of that particular population. But this means that someone must be found who is prepared to undertake the long and laborious task of measuring thousands of bones and putting the results into a computer. While it is probably not necessary to have a highly expert anatomist to perform this measuring operation it is certainly essential to find someone who will devote themselves to a tedious task.

Very recently, however, some new finds have caused a resurgence of interest in human skeletons and their measurements. A large cemetery of Romano-British times has been discovered at Poundbury, near Dorchester, with many interesting features such as the burial of fairly large numbers of people in lead coffins.

This has produced some interesting preservations, notably several entire heads of hair. It also allows studies on the prevalence of lead in skeletons, and may allow deductions to be made about contamination with lead during long periods of burial. This is interesting in view of present-day preoccupation with lead pollution and its possible long-term effects. There is also the prospect of discovering a large number of burials in the excavations recently started at York; anywhere between five and ten thousand bodies of many different periods are expected to emerge eventually from the York excavations. These discoveries allied to modern techniques in the laboratory may allow significant new results to be found. There are already hints that careful measurements are showing very small skeletal anomalies – small personal peculiarities in bone structure – and these are recurring in such numbers that there is a suspicion that they may be genetic. In this case some idea of family structure may emerge from these cemeteries. This resurgence of interest in skeletal measurement is typical of one or two interesting developments in this field of the study of man.

'Antigens' are those factors on the surface of individual cells which express the individuality of the genetic material inside the cell; in other words the antigens on the surface of the cells of one individual human differ from those on the surface of the cells of any other individual (unless the two are identical twins). The ability to recognize the antigens on the surface of a cell and to destroy all cells which are foreign – i.e. which do not carry its own personal set of antigens – is the basis of the body's power of resistance to infection. It is also the reason why humans 'reject' grafts or transplants from other people's bodies. The growth of transplant surgery has led to much closer study of this antigen-system and to the recognition that, while we are all different from each other, some of us are more alike than others, and in particular that there are a small number of these antigens which are closely involved in, and very powerful for, transplant rejection. These are called 'histocompatibility antigens' and they are studied in 'tissue-typing', the attempt to discover which donated organs are least likely to be rejected by the potential recipient of a transplant.

Recently anthropologists have taken up this work done by the immunologists for the benefit of the transplant surgeons. It is now claimed that different population groups, or races, have different 'antigen profiles', in the shape of different frequencies of the thirty known tissue-antigens, each individual having four of these antigens. It is further claimed that those populations which are most isolated, such as the Eskimos or the natives of New Guinea, have the most clearly defined antigen profiles.

Normally, in living people, antigens for tissue-typing are determined from blood samples, but recently Dr Peter Stastny of the Texas South-Western Medical School[4] has developed a method of determining antigen profiles from shredded tissue of mummies. Using this technique he has shown that thirty-five mummies from Peru present very similar antigen profiles to those of present-day South American Indians. This is in accordance with archaeological beliefs about the continuity of the present inhabitants from those of pre-Inca Peru. Stastny therefore believes that his tissue-typing techniques could be used to trace the movements of peoples in prehistoric times. He is now studying tissue from Egyptian mummies, but this line of investigation is necessarily limited to those few parts of the world in which mummies are preserved by natural, climatic dryness.

Yet so many similar hopes have been raised in the past, in vain. Many mummies have been X-rayed, and mummy-tissue has been examined by all sorts of different scientific techniques, but the information obtained has done little to supplement what is already known from Egyptian writing. The problem is that, where there are large numbers of well-preserved mummies, there are often also literary or documentary sources, or the peoples are near enough in time and space to such other forms of evidence that the study of the bodies reveals little that was not known already. In those places where there is little other evidence there is usually little organic tissue left – nothing more than the inorganic chemical framework of the bone. There have, of course, been outstanding finds of organic tissue – perhaps the most famous being the 'Grauballe' and 'Tollund' men in Danish bogs. In one case Hans

Helbaek was able even to analyse the contents of the stomach which contained a most extraordinary 'porridge' of plant seeds. But from only one individual body it is impossible to tell whether this was the normal diet of the people, or whether it was some special meal provided for a man about to die, to die perhaps for proven crime, perhaps in some type of sacrifice.

There is little doubt that the most successful study of man in archaeological terms has been the study of the disease-states of ancient peoples, sometimes called palaeopathology. This has provided evidence of tuberculosis in skeletons from Neolithic Germany, and osteoarthritis in a 45,000-year-old skeleton of a Neanderthaloid man from Shanidar in Iraq. Extraordinary bone-tumours have been found on the skulls of pre-Colombian Peruvian Indians and on the femur of a Fifth Dynasty Egyptian. There are also a number of well-known cases where skeletons show what appear to be the wounds caused by medical trepanning of the skull.

But while all these are interesting in themselves they tell us little except some detail about the otherwise unidentifiable human on whom the scars are found – unless it be to tell us that we are only heirs to the ills our ancestors suffered from. A more informative approach, sometimes called the epidemiological approach, has been developed largely by Don Brothwell. The pioneer work here was a study of the Pueblo Indians by Hooton as far back as 1930. This approach asks these questions about any earlier society:

> Did the community tolerate a heavy disease load? Which diseases may have prevailed and did any have special influence? What was the life expectancy? How high was infant mortality? Might there have been different disease patterns related to geographic background or social stratification? Could sanitation, habitation, diet, etc. have influenced community health? How did the community attempt to maintain and promote health or prevent disease (rational or ritual methods)? What of personal hygiene or public health practices? How did the group react to a disease (physical or ritual)? . . .

This quotation comes from Brothwell's paper entitled 'Community health as a factor in urban cultural evolution', and the title states clearly what the new approach tries to illuminate.

The facts about expectation of life and infant mortality can be found with reasonable accuracy, bearing in mind the qualification stated by other authors above, but only if a sufficiently large number of burials has been found to give a statistically valid sample for that community. But Brothwell's own early work on tuberculosis in ancient Egypt demonstrates only too clearly what has happened to much of the evidence. The conclusion of the study is that examination of thirty-one skeletons and mummies proved conclusively that tuberculosis did exist in ancient Egypt, although there are no conclusive artistic or literary references to the disease. But only sixteen of these cases had been reported by previous authors. Fifteen of them were entirely fresh studies, many of which were done with material which had been lying around in museums for years without anyone being aware of the evidence it could provide. Brothwell comments:

It is, alas, only too likely that the cases of possible tuberculosis reviewed in this paper represent only a part of those that have actually been found at Egyptian sites . . . How many specimens in a fragmentary condition have been discarded, as so much valuable skeletal material has been in the past? . . . Because of these problems of past excavations, it would be hazardous to try to estimate the frequency of spinal tuberculosis in earlier Egyptian populations. From the evidence it seems likely that tuberculosis was established by pre-Dynastic times and was certainly present by very early dynastic periods.[5]

Much of the evidence that would have given us the power to make studies of large samples of prehistoric men and their health has, therefore, disappeared, because archaeologists of earlier generations have not realized how much can be obtained from the examination of skeletons. Even when large finds of skeletal material are made nowadays – and they are becoming increasingly rare – and even though we now have better methods of preserving bone, there are pitifully few people willing or able to examine the finds from a pathologist's point of view.

It would seem, though, that we ought to be able, with modern scientific methods, to find out a good deal about the diet of prehistoric men. The work on coprolites at Tehuacan, by Callen, mentioned in Chapter 10, is an outstanding example, where even some account of social habits could be deduced from the evidence. The stomach contents of Grauballe Man, mentioned above, are yet another example of the sort of evidence we might hope to find and interpret. But these examples seem to stand out in rather lonely isolation from the general achievements in this area. Skeletons, which are the most common direct evidence of the bodies of prehistoric men, can only show traces of those diseases where inadequate diet has been the cause; they tell us nothing about the causes of inadequate diet.

The disease of rickets, caused by deficiency in vitamin D intake, is marked by bowed bones. In life, sufferers from rickets often have grotesquely bowed legs, and this was a common disease of industrial slums in the nineteenth and early twentieth centuries – indeed, in the author's youth it was common to see among the older generation of Yorkshire industrial workers the traces of the disease. Some clear examples of rickets have been found in Egyptian skeletons, but the number of recorded cases is surprisingly few. Brothwell suggests that this is because the early Egyptians ate a good deal of fish from the River Nile and got plenty of sunshine. Rickets has appeared more often in finds from Scandinavia, and has also been found to be prevalent among Neolithic skeletons in Denmark. This geographical distribution is precisely what modern medical opinion would expect, and adds little to our knowledge.

Dental decay and bad teeth are clearly shown in the archaeological record, for teeth are among the commonest of all finds. Man's earliest ancestors seem to have suffered in this way, for dental caries can be clearly shown in the remains of 'Rhodesia Man', who long preceded *Homo sapiens*. One of the most outstanding achievements of palaeopathology has been in this field, where Brothwell has shown the steadily increasing incidence of dental caries among the British, French and Greek populations from Neolithic times to the present day. There is only one major variation in the curve, and that comes at the time of the collapse of the Roman Empire. The harder-living barbarians had better teeth than the civilized people they conquered, but as the Dark Ages gave way to the softer influences of civilization again, so the incidence of dental caries increased again, and continues to increase steadily up to our own times. But this, of course, parallels studies among different types of populations living now; broadly speaking, the coarser the diet, the less dental caries.

This slight disappointment that one can sense over the results of studying skeletons with the eyes of a modern pathologist really boils down to saying that no study of this type of material has yet revealed anything of importance that we could not have guessed on the basis of our general knowledge of man's health and diet.

There has, too, been one major disappointment in this area. Considerable hope was placed, at one time, in the study of 'Harris's lines' in skeletons. Harris's lines are zones of calcified material which are found in most bones. It was thought that they were brought into existence whenever the normal processes of growth were brought to a temporary halt. There should therefore be Harris's lines in the bones corresponding to childhood diseases and to periods of malnutrition, or to any other cause that would bring bodily development to a halt. These lines were, in fact, found in skeletons as far back as Neanderthal Man. They have also been found in Egyptian mummies, in Iron Age skeletons and in the bones from medieval cemeteries. But on closer study and with further work on contemporary examples by modern research pathologists, it has become more and more difficult to make clear interpretations of the lines themselves or of their possible meanings. Brothwell writes, 'Further studies are needed to equate this interesting anomaly with seasonal malnutrition and periodic famine. This may be more difficult than was at first thought, for recent work has shown that the association between these calcified zones and poor health in childhood is by no means clear-cut.'

The unique ten-year-long series of excavations in the historic British city of Winchester, which many consider to be the real start of a new sub-discipline of 'urban archaeology', are providing interesting materials to extend the study of biological man. (The Winchester excavations will be dealt with more fully in Chapter 17.) The large number of skeletons unearthed from cemeteries extending from Saxon to late medieval times is being studied by Brothwell, and this alone represents a colossal task. It is already clear that a very high proportion of these skeletons show traces of arthritis, which will probably surprise no one who knows the British climate, but is at least a scientific confirmation of our 'folk-knowledge'. They also demonstrate medieval medicine, in the shape of mended fractures, and they show tumours both tubercular and cancerous.

One of the most interesting observations is that some of the long-bone surfaces of medieval skeletons dating to somewhere before 1500, show a roughening which could have been caused by syphilis or allied diseases such as yaws. This throws light on the old argument as to whether syphilis originated in the New World and was brought back to Europe by Spanish conquistadors, or whether Europe had syphilis in its own right. Brothwell has pointed out that there is evidence of a disease at least like syphilis in an edict of Edward III warning the mayor and sheriffs of London against allowing people to indulge in 'carnal intercourse with women in stews and other secret places', because of the danger of contracting the disease. But again we get an obfuscation of the issue in a recent suggestion that all diseases caused by treponemes (the agents of syphilis and yaws) may be caused by one basic organism in different variants, and it is only the results of competition between these variants that decides whether a population group suffers from yaws or syphilis.

Certainly the medieval population of Winchester must have suffered gravely from intestinal worms and flukes, for excavation of medieval cesspits revealed large numbers of eggs of at least three species of these parasites on humans. Although the eggs were dead, they were wonderfully preserved, and easily recognized by two scientists from the Commonwealth Bureau of Helminthology, Dr H. H. Williams and A. W. Pike. Even the embryo within the egg was visible in some cases.

This remarkable success led to discussion of the possibility that micro-organisms might be found directly. Sure enough, a little while later, when this possibility was in the excavators' minds it was found possible to recover samples of the tetanus germ, a bacillus, from a fifteenth-century drain. The problem here lay, not in identification, but in proving that the germ itself had been medieval; it was always possible that the micro-organisms had been carried down to the level of the drain by water percolation in later years, perhaps even in recent days. Because it seems quite impossible to guarantee that there has been no contamination from the surface in these cases the problem of how long bacteria remain viable will be difficult to solve.

Another technique is now being tried to see if traces can be found of bacteria which are long since dead. The toughest part of a bacterium is, naturally, its cell-wall, and this cell-wall carries the antigens which distinguish the bacterium. (Bacterial antigens are similar to the human antigens described above, except that individual bacteria do not have individual antigens, and the antigens of one diphtheria germ are the same as those of millions of others of the same type.) In day-to-day clinical medicine of our own time these antigens on bacteria are used regularly for identifying the type of infections from which patients are suffering. The technique is to take samples from the patient and see which of a selection of previously prepared antibodies will destroy the germs. This is a highly selective and discriminating method because the essence of antibody is that it will attack only the type of germ which has the particular antigens which correspond to the shape and chemical structure of the antibody. A variation of this technique is to attach a fluorescing chemical to the antibody, so that any grouping of antibodies, attacking a germ by attaching themselves to it, can easily be seen under the microscope. This technique is now being used on samples from the Winchester cesspits in the hope that modern antibodies will still recognize the antigens that may emain on the tough outer skins of long-dead medieval bacteria.

But the difficulty with all these studies on human remains in the archaeological context is that they seem so scattered, so specific to each individual site or excavation, sometimes specific to just one single skeleton. Nowhere, except as a promise in Brothwell's epidemiological approach, does any generalization occur. Quite possibly this is because of lack of manpower in this field. It may be, however, the lack of one man with the qualifications and enthusiasm to pull many disparate pieces of evidence together.

This difficulty is well illustrated by Marcus S. Goldstein, of the Department of Health, Education and Welfare in Washington.

There . . . is a definite need for compiling the bits of information on palaeopathology that are often 'buried' in archaeological reports and the like. In many instances the skeletal remains uncovered in an archaeological excavation are few in number, and hence a report on them, including observations on pathology, appears as a brief addendum to the archaeological monographs. Such fragmentary data on skeletal remains and palaeopathology may or may not be significant in any single instance; when multiplied a hundredfold they might well reveal much of interest to a large circle of workers. It is suggested therefore, that 'clearing-houses' be established at places willing to assume the responsibility, to which could be sent all publications containing data on palaeopathology.[6]

Until some such scheme appears and produces fruit, it seems likely that we shall continue to know more about our ancestors' artefacts than about their arthritis; we have certainly at the moment more information about their burial customs than about the causes of their deaths.

1. D. Brothwell and E. Higgs, 'Introduction' in *Science in Archaeology*.
2. C. Coon, *The Origin of Races*, p. ix.
3. L. H. Wells, *Stature in Earlier Races of Mankind*'.
4. The work of Peter Stastny has been widely reported, notably in *Newsweek*, 13 November 1972.
5. D. Brothwell *et al.*, 'Tuberculosis in Ancient Egypt'.
6. M. Goldstein, 'The Palaeopathology of Human Skeletal Remains'.

13
Fakes

Archaeology has been troubled by fakes and pestered by fakers ever since it began. The workmen employed by Boucher de Perthes to dig in the gravel pits of the Somme Valley, the very men who extracted from the earth the first convincing proofs that their species was of vast antiquity, learned the value of their work very quickly. They soon discovered how to 'salt' a new excavation with stone axes which they had, in fact, picked up at one of the earlier digs. But these first forgers were not very skilful and their activities were soon revealed, and revealed by those who most opposed the theories that Boucher de Perthes was putting forward. So poor Boucher de Perthes had to face attacks on his integrity and on the credibility of his whole work, as well as to outface the vast majority of savants of his time with his revolutionary ideas about the antiquity of man. It took him years of arguing and publishing to establish the genuineness of his findings.

But archaeology is much more than a science. The excavator is the main source of supply to the displays of most of the world's great museums. Archaeology therefore shares in the excitement of a treasure hunt, and the flavour of the tomb-robber's sacrilege is in it, too. The scholar may say, and truthfully, that none of this enters his mind or his motivation, but these excitements and guilts are there for the public, with whom he must eventually communicate if his work is to be anything but esoteric. This glamour automatically starts giving a financial and emotional value to the finds of excavation.

Archaeology also has strong connections with art. Art-historians were the 'experts' of archaeology before the laboratory scientist found he had a place in the subject; it was only by their criteria that many problems of relative dating or the inter-connections between different cultures could be solved. The objects found and studied in this manner tended, therefore, to become *objets d'art* in the strict meaning of that phrase. Some of them were, of course, genuinely beautiful; many were made of metals and materials we still consider to be precious or valuable. Thus the monetary values of the saleroom became applicable to some of the data of the study of man's history, and archaeology was dragged into that most repellent area of our modern culture where works of art become 'investments' of the affluent and acquisitive. It is widely believed in Europe (though I have never had first-hand proof) that many American collectors of great wealth have stolen masterpieces and smuggled antiquities which they keep for their own private pleasure. What is certainly true is that smuggling goes on, and museums are then placed in a difficult position: either they buy what they are offered or they face the risk that important works will disappear for ever into an underworld of illicit private ownership. The loss to scholarship and history will then be enormous, for not only will the piece be unavailable for study, but its value as evidence will be reduced, perhaps to nothing, as its pedigree (its provenance in the technical jargon) the precise detail of where and how it was discovered or excavated, will be more than lost – it must be hidden to protect the owner.

In this shady world where unusual or valuable or ancient objects mysteriously 'come on the market', the forger and the faker come into their own. There can be little sympathy for the man or the museum defrauded in a deal they know to be illicit, but there seems no reason why museums should not accept the 'Philadelphia Declaration', a self-denying ordinance which proposes that all museums should accept a code of ethics and refuse to buy any piece for which there was no full 'provenance'. This would cut the forgers' market by half.

Yet probably every major museum in the world has an embarrassing chapter in its history, an object locked away from public view if only to hide blushes, where it has been 'taken for a ride' by a forger intent on financial gain. One of the most famous 'misjudgements' was the purchase by the French authorities (with wide academic support it must be said), of the 'Tiara of Saitapharnes'. This was supposed to be a superb example of Scythian goldwork, but turned out to be a product of nineteenth-century Odessa. The Metropolitan in New York have had an even more trying experience, more recently, when an ancient Greek horse, undoubtedly one of the loveliest objects ever to appear in any museum, was clearly shown to have been made by methods quite unknown to the Greeks. However, a further study of the object in 1972 showed that while definitely a forgery, it was undoubtedly ancient, and the final conclusion is that it is the finest example known of ancient forgery – almost certainly a Roman fake of some genuine Greek work of art.

Undoubtedly one of the greatest benefits that laboratory science brings to the archaeologist is the opportunity to establish, and usually to establish with absolute certainty, the genuineness of a discovery. In practice, the entry of the laboratory scientist into the archaeological field has often resulted from the problems of conservation of museum specimens. In learning how to preserve items of our heritage against the ravages of time, temperature, exposure to light and the disintegrating effects of the polluted atmosphere of the twentieth century, scientists have been forced to learn the history of the technology of art. Now this knowledge of ancient techniques can be used to expose the forger and the fake.

Until about 1950 a major item in a museum could only be shown to be a fake, or even a genuine mistake, by a slow building-up of a consensus of academic and scholarly opinion, based on typologies, on art-historical criticism, on the relative dating of the object in relation to possibly contemporary cultures, and on a general ability to fit the object into some rational over-all picture of the past. Since there might well be strong political or nationalistic reasons for accepting an object as genuine – as there were in the case of some of the German fakes that provided support for the Nazi myths of the 1930s – it could be almost impossible for unbiased scholarly opinion to prevail.

But the museum laboratory provided the new techniques to settle the matter of genuineness, in a sharp and scientific way. The British Museum Laboratory, to take one instance, has recently celebrated its fiftieth anniversary. It was set up in 1922, because many of the museum's treasures had been stored in an unused section of London's underground railway during the First World War, and they had deteriorated so much that it was clear that new forms of remedial treatment were required to save many of them. A single scientist (Dr Alexander Scott), with two assistants, started a temporary laboratory, originally designed to exist only for three years. Research into matters like corrosion brought knowledge of the chemical composition of the bronzes used by Greek sculptors, of the purity of the gold used in Merovingian coins, of the nature of the pigment used by Renaissance painters. This knowledge, it turned out, could be used not only to preserve the ancient works of art, but also to show that pieces of metalwork, allegedly Greek or Roman, could not possibly be so, because they contained, as part of the chemical composition of the alloy, a metal like zinc which had been unknown to the technology of age in which they were supposed to have been made.

This step into scientific analysis of materials in order to establish the genuineness of a disputed object seems obvious to us nowadays. It was not taken, however, as soon as it might have been, save in a very few exceptional cases. The reason for this is clear: a museum holding some unique and extremely valuable object, an object that is priceless for either scientific or artistic reasons, will be most unwilling to let the scientist handle it in the way which is necessary to carry out most tests. The shape of the object clearly must not be changed; bits cannot be cut off or bored out without damaging it in a way that many will find peculiarly offensive, namely that the object will bear the scars of the impact of modern technology and will be no longer what it was, the unchanged, total concept that came from a master's hand. There has, therefore, been a premium placed on the development of non-destructive testing methods, such as the use of X-ray or ultra-violet exposures. Simply examining an object under X-rays or by ultra-violet light will not of itself reveal whether it is a fake. So there has had to be further research to discover what are the characteristics of genuine objects under these forms of examination, in order that the deceptions of the faker may be exposed. The clearest example of success here has been in obtaining knowledge of ancient techniques of casting metal objects, where the more efficient methods of modern technology show up under X-rays when the forger has produced the article.

Faking and forging, however, are not necessarily performed for financial gain. The motives may be more complex, and therefore the fakes may be harder to discover. A misdirected chauvinism, or a distorted racial pride seem to have been the driving forces, not only in the Nazi fakes of the 1930s, but in such 'discoveries' as the 'Minnesota Stone' in the Midwest of the U.S.A., where apparently Runic inscriptions implied that the Vikings had penetrated to the interior of the American continent. The French 'Gloziel' tablets were in a similar category, apparently, seeming to prove that there had been a literate civilization in France many thousands of years before the first civilizations arose in the Middle East. But in these cases the fraudulence was so obvious that it did not need scientific methods or laboratory tests to expose them.

Much more interesting is the case of the Piltdown Skull. The basic motive here seems to have been forgery in the cause of promoting a scientific theory, but there may well have been an admixture of British chauvinism too. Protection of the fake by the natural caution of museum authorities played a role in delaying the exposure of the fraud, but even more protection was granted by the scientific climate of opinion, which favoured those theories supported by the existence of the skull. Indeed it was only when scientific opinion swung against those theories that the techniques of laboratory analysis were allowed to be performed on the bones that formed the data. The whole story of Piltdown has all the excitement of a detective novel with the added spice that we still do not know who was the true villain, although in the last few months the burden of suspicion has shifted dramatically from the most obvious suspect and there are hints of some mysterious figure in the background, while at the same time one of the minor characters seems to have been pretty well cleared from the hunt.

The essence of the Piltdown find was that a skull, fragmented but of the same type and size as that owned by modern *Homo sapiens*, was found beside a jawbone which had definitely ape-like characteristics. There were also three teeth which fitted the jawbone, one of which was basically ape-like in shape, yet appeared by the flattening at the top to have been used to consume a diet similar to that of man. Together with these exciting remains were found remains of animals, notably rhinoceros and a type of elephant, which were known to have existed half a million years ago, long before the end of the Ice Ages.

These finds, made in 1912, fitted exactly into the theories that most scientists then held about the origins of man; in the simple form of Darwinian evolutionism then prevalent, man had arisen from the apes primarily because of an expansion of his brain, and a creature like Piltdown Man *should* have existed. The fact that he had been found surrounded by some primitive stone tools and a curious bone object and the remains of the animals he had probably slaughtered was an extra bonus. The fact that he had lived on the Downs of southern England was a source of pride.

The first finds, chocolate-brown-coloured parts of a remarkably thick human skull, with fossilized hippopotamus and elephant teeth and crude flint tools, were brought to Smith Woodward, the Keeper of Geology at the Natural History Museum in London, by Charles Dawson, a country solicitor and keen amateur archaeologist. He reported that workmen had found them near the village of Piltdown in Sussex, while digging gravel from deposits believed to date from the beginning of the Ice Age. Woodward agreed to join Dawson in a further dig and they then found fragments of the ape-like jawbone with two teeth, more pieces of skull-bone, very human in appearance, and more fossilized animal remains and more flint tools.

From the very start there was controversy. Woodward claimed that the jawbone and the skull belonged to the same individual, and that this must be a new type of creature evolving from ape towards man. He called him Dawn Man, Eoanthropus Dawsonii. It was, of course, possible to argue that the jaw of an ape and the skull of a man had got into the same deposit by coincidence. But the fact that both jaw and skull were of the same colour and appeared to be equally fossilized and the unlikelihood of the coincidence justified Woodward in his conclusion on the basis of the evidence and the current theories of the time. In the following year a canine tooth was found at the same site; this appeared to match the jaw. It was the large protruding type of canine found in modern chimpanzees, but it was worn in a way never found among modern apes, though similar to the type of wear found in human teeth. Woodward still had opponents – David Waterston, a professor at King's College, London, the American zoologist, Gerritt S. Miller, of the Smithsonian Institute in Washington, and Professor Boule from France – but the evidence for Piltdown Man was strong enough to carry the consensus of scientific opinion in favour of Woodward's interpretation.

The case for Piltdown Man appeared complete when the remains of a second individual were found in 1915 about two miles from the original site – more pieces of human-like skull with an ape-like molar tooth. Soon Gerritt Miller was the only major opponent of international status and he, as early as 1915, wrote, 'Deliberate malice could hardly have been more successful than the hazards of deposition in so breaking the fossils as to give free scope to individual judgement in fitting the parts together.' Miller's argument was that that jawbone could never have been attached to that skull by any conceivable anatomical arrangement, so he allowed Woodward's description of the skull as that of a previously unknown fossil man, but he claimed the jaw belonged to a previously unknown type of fossil chimpanzee. (By 1930 Miller believed that there was fraud involved but his colleagues persuaded him not to say so publicly without positive proof.) Meanwhile the remains of Piltdown Man were housed in the Natural History Museum as one of the most precious of exhibits and visiting scientists were rarely allowed even to handle them.

It was not the work of laboratory scientists, at least in the first instance, that cast doubt upon Piltdown Man; it was the work of excavators in many different parts of the world. Pekin Man was discovered in China, further examples of Java Man made up a rather different picture, and in South Africa the first examples of Australopithecus, a hominid, were found. A very different theory of the evolution of man began to hold the field, a theory which held that man evolved from some ancestor common to both himself and the apes, and that the distinctively large brain was the final stage of evolution. Piltdown Man did not fit into this genealogical tree; he became an anomaly, even an embarrassment, until finally only his great age allowed him to be accepted as possible at all.

This situation had developed by the end of the Second World War, and during that war a young scientist from the Natural History Museum, Kenneth P. Oakley, had been drafted into the Geological Survey. Oakley was an expert on fossil invertebrates but anthropology was his hobby, and he helped the Museum to mount a wartime exhibition called 'Man, the Toolmaker', which brought him face to face with the problems posed by Piltdown Man. At the same time his wartime work made him a colleague of a geologist who was working on the problem of fluorosis, a disease caused by the amount of fluorine in the diet. It was at a meeting in 1943, concerned with the fluorosis field, that it was suggested to Oakley that the amount of fluorine in a skeleton could provide a method of relative dating. In fact the idea had first been proposed by a French scientist, Carnot, in the 1890s, and, although he had made little progress in the work, the suggestion had been kept alive by a French author, Vayson de Pradenne, in a book on archaeological frauds.

Fluorine is present in virtually all soils and in the ground-water in those soils, though the amount of fluorine in one type of soil will differ from the amount in another type. For strictly chemical reasons fluorine atoms from the soil and the ground-water are attracted into buried bone where they replace atoms that are part of the structure of bone when it is alive. This process is a steady one dependent on time and the amount of fluorine in the soil.

Bones buried at the same time in one type of soil at one site should therefore show the same amount of fluorine in them. Bones buried at a different site in a different soil will show a different amount of fluorine, even though they were buried at the same time as the first lot. Fluorine therefore cannot give an absolute date to buried bone, but it can tell whether two bones buried in the same place have been buried for the same length of time. The incorporation of fluorine into bone is not the same as fossilization, which is essentially the displacement of material which was originally organic by material which is purely mineral. This fossilization process allows another method of relative dating of bone, by the measurement of the amount of nitrogen in the bone. The amount of nitrogen is a measure of the amount of organic material left in bone, and is therefore a clock telling the amount of fossilization that has taken place. But it is not a satisfactory universal clock because it has been discovered that the soil conditions at any place will speed up or slow down the removal of nitrogen from bone. But in general bones buried in the same place at the same time will have the same amount of nitrogen in them, just as they will have the same amount of fluorine in them. The longer they have lain in the soil the more fluorine and the less nitrogen they will have.

When Oakley returned to the Natural History Museum at the end of the war he was encouraged to continue his interest in fossil man, shown in his helping with the 'Man the Toolmaker' exhibition, and in 1948 he was asked for his advice in connection with the Galley Hill skeleton. This was a skeleton which had been dug out of gravel beds on the south bank of the Thames estuary which were more or less contemporary with the stratum that had contained the ancient, genuine and famous Swanscombe skull. The Galley Hill skeleton had been dug up in 1888 under not very scientific conditions, with early stone axes and elephant remains around it. It was clearly *Homo sapiens* yet the gravels in which it lay could be dated to 100,000 years old, which would have made *Homo sapiens* precede Neanderthal Man. Galley Hill had always been controversial and now, in 1948, there arose the possibility of the skeleton being sold. For the first time the fluorine and nitrogen tests were used. They showed clearly that the Swanscombe skull and bones of fossil mammals from the gravel at the site contained more than 1½ per cent of flourine and virtually no nitrogen. The Galley Hill skeleton showed only a third as much fluorine but more than 1½ per cent nitrogen – roughly the same measurements as those given by a known Neolithic skull from a near-by site. The Galley Hill skeleton was therefore that of a man buried in the gravel ritually whose remains had, by chance, been laid in a grave dug down to the level containing fossil remains 100,000 years old – in technical terms an intrusive burial.

The successful resolution of the Galley Hill problem and the increasingly embarrassing divergence of Piltdown Man from all current theories persuaded the Museum authorities to allow Oakley to examine the Piltdown material and to take very small samples for chemical analysis. By this time, 1949, the Piltdown material consisted of the jawbone with teeth, the canine tooth that was part of the 'second individual', parts of human skull from both first and second sites and seventeen fossil remains of mammals from both sites. These animal remains consisted of two groups, one containing remains of elephant and possibly hippopotamus which could be roughly dated back to about 600,000 years, which is at the boundary between what geologists call the Pliocene and the Pleistocene Ages, and the other containing remains of animals like beaver and red deer which could be dated to about 60,000 years ago – just before the start of the last southward expansion of the ice sheets. It appeared that these two groups of fossils had been brought together into one deposit, a former river bed, by sheer chance.

The fluorine analysis was performed by Dr C. R. Hoskins at the Government Chemists' Laboratory. The results came as an enormous shock. They showed that Piltdown Man was nothing like half a million years old. He was more likely to be about 50,000 years old, dating back only to the later part of the Ice Age. In detail the figures showed that the animal fossils known to be of great age, notably the elephant teeth, contained virtually no nitrogen, but a great deal of fluorine, sometimes as much as 3 per cent. The Piltdown skull bones contained, on the other hand, 1.4 per cent nitrogen and barely 0.1 per cent fluorine. The jawbone contained even less fluorine and even more nitrogen, and the canine tooth was similar.

In evolutionary terms, therefore, Piltdown Man was an absurdity, a creature with no known ancestors and no known descendants. He could not be the ancestor of man, because man already lived 50,000 years ago. But how could he be ape when no other ape fossil was known from Europe for the last million years? Yet the state of scientific knowledge after the fluorine tests could not be pressed any further than the statement that Piltdown Man had lived in the Ice Age at the earliest. There was no proof at all that there was any forgery involved and the warning given by Gerritt Miller so many years before had been virtually forgotten. The chemical tests still allowed Piltdown Man to exist, in the sense that the skull and jawbone seemed to be of the same age, and there was also the confusing factor of the hippopotamus tooth, which was clearly fossilized and which seemed to contain about the same amount of fluorine as the 'human' remains, though how a hippopotamus came to live in Britain during the cold of the final stages of the Ice Age was even more of a mystery.

During the investigation of the Galley Hill skeleton, however, Oakley had been brought into contact with the anatomy school at Oxford under Professor Sir Wilfred le Gros Clark, and in particular with J. S. Weiner, who followed the fluorine tests on both Galley Hill and Piltdown remains with great interest. It was Weiner who moved the affair into the next steps. Brooding over the absurdity of Dawn Man he came to realize that the teeth were the only real connection between the humanity of the skull and the ape-likeness of the jawbone. He started investigating modern chimpanzee teeth, and he tried grinding and filing some specimens to see if he could reproduce the appearance of Piltdown Man's teeth. He was successful to outward appearances though examination under powerful microscopes showed the regularity of filemarks on the ground-down surfaces. It was in 1953 that Weiner took his results officially to Professor le Gros Clark (who had not then been knighted). Professor Clark had apparently become suspicious independently, for he was planning a trip to Sussex to re-examine Dawson's area of operations. They went jointly to Oakley and suggested seriously that Piltdown had been a fraud. It was decided to join forces, to bring all possible techniques to bear, and to see if it could be proved scientifically that there had been forgery.

By this time methods of chemical analysis had been, quite independently, made much more discriminating. Also by this time the Museum authorities were willing to let larger samples of material be taken from the Piltdown remains. Suddenly the evidence poured in. The analysis at the Government Chemists' Laboratory, using the finer methods developed by C. F. M. Fryd, showed that the jawbone and teeth contained no more fluorine than modern bone, whereas the skull-bone contained just enough fluorine to imply that it was genuinely ancient, though the fluorine level was lower than that of any bone from Pleistocene gravel anywhere else in Britain. The nitrogen content of jawbone and teeth was likewise the equivalent of that in modern bones.

The first close and scientific examination of the teeth showed they had been flattened artificially; there were the straight criss-cross marks of some modern abrasive revealed by high-powered microscopy, and the teeth were now seen to have been levelled down evenly all over the wearing surfaces, completely unlike natural wear. X-ray pictures showed that the X-rays of 1913 had misled scientists for forty years. The early X-rays had shown short roots to the teeth as in modern men's teeth; the X-rays of 1953 showed typically long, ape-like roots. Radiography revealed that the grinding had at one point penetrated to the pulp-cavity of the canine tooth and that this hole had been filled in with radio-opaque material.

Even more damning was the evidence that came from the National Gallery, which has a small scientific staff. They proved that the brown stain on the canine tooth was a pigment, probably Vandyke Brown. So tests were started on the colouring of the skull bones, which were admittedly genuinely old. Gypsum was found as a chemical in the skull bones, yet there is no gypsum in the soil at Piltdown. However, an acid iron-sulphate solution used on partly fossilized bone causes a reaction which produces gypsum, and stains the bone a chocolate brown. Almost certainly this is what was done. The electron microscope also added its quota to the damning evidence; it revealed perfectly preserved portions of organic tissue in the jawbone, proving it was modern, while no trace of such material could be found in the skull-bone.

The mystery of the hippopotamus tooth found its solution in independent work by other archaeologists. The limestone caves of Malta have been found to contain large quantities of fossils, including samples of hippopotamus teeth, and in the peculiar conditions of limestone deposits it turns out that there is a very low rate of fluorine-transfer to bone. There is no proof that the Piltdown hippopotamus tooth came from Malta – similar conditions have since been found to obtain in British limestone caves – but the fluorine content of the tooth is of the same order as that found in Maltese specimens.

There was also a tooth of a very rare, extinct form of elephant found at Piltdown. This was analysed, along with many other specimens, by S. H. U. Bowie, for uranium content. Uranium atoms transfer into buried bone just like fluorine atoms, and since they are radioactive, the amount of uranium in a bone can be measured without destroying the specimen. The uranium content of the elephant tooth was far higher than the uranium found in any other British fossil of Pleistocene or Tertiary times, and higher than that of any elephant teeth tested from anywhere else in the world so far, with the exception of one site only. This solitary site is at Ichkeul in Tunisia, where one example of an elephant called *Archidiskodon africanavus* has been excavated, and these remains show exactly the same fluorine content and virtually the same uranium content as the tooth found at Piltdown. This was not final proof of the origin of the Piltdown tooth but it was very suggestive, once it had been shown that skull and jaw were of different ages and had been coloured to match.

The actual preparation of the samples that provided this evidence had also proved suggestive to Oakley. When he was drilling his sample from the jawbone of Piltdown Man the whole of the room was filled with the distinctive and unpleasant odour of burning horn as the drill burr bit deep and started warming up. Furthermore, the material came out of the drill-hole looking like shavings. The collection of the sample from the skull-bone produced neither smell nor shavings, and the conclusion was clearly that the jawbone was modern bone, while the skull-bone was a genuine fossil.

It was in November 1953 that Oakley and Weiner with the backing of le Gros Clark published their results and declared Piltdown Man officially a fraud. The jawbone and teeth were those of a modern orang-utang, stained, ground down and thoroughly faked. The skull was genuinely ancient but had been treated with stain and with chemicals which produced an effect like fossilization. The most significant of the remains of mammals found with the human remains were genuine fossils, but had been introduced from other sites after having been stained in a way which pronounced their introduction to the site as deliberate fraud. The 'second individual' of Piltdown Man consisted of parts of the same material which had not been used in the first 'find'. It was possible that the skull bones were from an individual of *Homo sapiens* suffering from a pathological condition which thickened the bones. The scientific problem of Piltdown Man had been solved by showing that he had never existed.

But the human problem remains with us, unsolved. The reason why so many – the vast majority of – scientists accepted Piltdown Man for so long was primarily because the climate of scientific opinion favoured the existence of such a creature. But who perpetrated the fake and why? That Dawson himself planted the evidence seems reasonably well assured. In the last few weeks before writing it has been revealed in the June 1973 issue of *Antiquity* that another major find of Dawson's was fraudulent. These were inscribed Roman bricks from the Fort of Pevensey (also in Sussex), and were the only known archaeological evidence to support the claim made in the writings of Claudian that the Roman general Stilicho, under the Emperor Honorius, had defeated the Irish and Picts and refurbished the forts of the Saxon shore between A.D. 395 and 399. But now chemical analysis of the bricks has shown that they are not similar to the other bricks in the undoubtedly Roman remains of Pevensey Fort, and thermoluminescence dating shows that bricks were fired almost certainly within the present century.

These bricks were 'found' by Dawson in 1902 and exhibited before the Society of Antiquaries in London in 1907. One of them is in the British Museum, and there is some confusion over whether there were originally four or three specimens. Their importance in historical terms is small, though they have been given some mention in academic discussion of the precise date of the end of the Roman occupation of Britain.

But the evidence from the laboratory 'leaves little room for doubt that the (brick) stamps were forged in the early twentieth century', according to D. P. S. Peacock, the author of the article in *Antiquity*. He concludes with the suggestion that 'the time is now ripe for a full investigation of Dawson's numerous and often bizarre discoveries', particularly some cast-iron figurines reputed to come from the Roman ironworking site at Beaufort Park. The forged brick-stamps also have implications regarding the Piltdown affair; Peacock says, 'It now seems that Dawson was associated with another fraud at a date prior to the Piltdown affair, which firmly points the finger of suspicion in his direction, although there is of course no proof that he was not the innocent dupe of another party on both occasions.'

Some people have favoured Père Teilhard de Chardin as the author of the Piltdown hoax. It is easy to see why, since the Jesuit priest is an extremely unpopular figure among many scientists on account of his theories of creation and evolution which attempt to reconcile science and religion. Teilhard de Chardin was a distinguished palaeologist in his own right, and he was present at Piltdown at the time some of the finds were made in his capacity as an interested scientist. Yet among all his writings he never mentioned Piltdown or the evidence it provided; he argued neither for or against, he ignored it, though he is reputed to have said

'How odd that it should be found in England,' or words to that effect. Those who favour his guilt claim his silence as evidence against him. It is equally arguable that his silence came from suspicion about the genuineness of the finds. But the chronology of his visit to Piltdown makes it virtually impossible that he could have prepared and performed such an elaborate hoax. The new evidence that Dawson was involved in another hoax years before he met Teilhard de Chardin (in 1908) must also weigh heavily against the Jesuit's involvement in Piltdown in any guilty way.

But if the piling up of evidence tends to remove suspicion from Teilhard de Chardin it makes the case against Dawson more complex. Dawson was, after all, only a country solicitor. How could he have prepared faked material of so many different kinds so skilfully? Where could he have got the knowledge of staining and fossilization so as to deceive the most eminent scientists of his day? How could he have accumulated the scholarship to know just what fakes would interest historians who specialized in later Roman Britain, and zoologists and anatomists who interested themselves in evolution, especially the early evolution of man? How could he have got hold of the raw material for the Piltdown hoax? A country solicitor, even if a keen amateur archaeologist, would be hard put to find a modern orang-utang jaw, the fossilized skull of a diseased Neolithic man, a hippopotamus tooth from a Maltese cave and the remains of a rare elephant from Tunisia. The implication is that there was a scholar with access to large collection of exotic material behind Dawson, possibly even using him as an innocent tool.

The importance of Piltdown in the context of this book is that it was a 'scientific' fraud – a fraud to prove a scientific theory. It was eventually exposed by scientific methods, and archaeology played only the role of handmaiden to the whole story. The only other frauds considered here in detail are essentially 'archaeological', that is to say the value placed on the fraudulent material arises primarily from archaeological activity. The exposure of the frauds has been achieved by a scientific technique which was originally devised specifically to help the archaeologist – the technique of thermoluminescent dating (described in detail in

Chapter 7). It is therefore necessary to start the story of these fakes with the archaeological background.

In a series of excavations in south-western Turkey from 1957 to 1970, James Mellaart, of the British Institute for Archaeology at Ankara, now a lecturer at the Institute of Archaeology in London, has shown that the Anatolian Plateau must be considered as one of the nuclear areas of Middle Eastern civilization, just as important as the Zagros mountains or any other part of the Fertile Crescent. In particular, he has shown that the enormous 'tell' of Cetal Huyuk conceals the remains of what is undoubtedly the earliest 'town' to be discovered anywhere in the world, a Neolithic town of a magnificence and sophistication that no one had suspected the earliest farmers to be capable of reaching. The greatest feature of this town is the impressive series of temples or shrines, containing furnishings and wall-paintings, many of them based on arrays of wild-bull horns which surpass anything found at any other site of comparably early date. Yet in their very first stages the people of this part of the world show marked links with the first inhabitants of Jericho, and the dates for both places start at about the same level – that is around 7000 B.C.

Mellaart was led to this part of the world by the discovery, by villagers, at a place called Hacilar, two hundred miles west of Catal Huyuk, of extraordinary painted pottery. Some of the digging had undoubtedly been clandestine, but many of the pieces found their way into reputable hands and were later included in Mellaart's official publication of the Hacilar site. For three years, from 1958 to 1960, Mellaart excavated Hacilar and showed that the site started on virgin soil with the arrival of a group of primitive farmers who had not yet discovered the use of pottery. The foundation of the site was dated to 7000 B.C., and the history of the place went through the normal sequence of desertions, burnings, rebuildings, development of local traditions, arrival of new peoples with different traditions and so on. But throughout the millennium from 5750 B.C. to rather after 5000 B.C. there was a remarkable pottery tradition, with coloured pots, sometimes painted, and a succession of beautiful, if primitive, female figurines, which first of all develop in style and then deteriorate. In the later stages of the culture there appeared some very distinctive pots modelled on human shapes. Sometimes there was one head, sometimes two; often small pieces of obsidian, inlaid in the pottery, marked the eyes; many were painted with simple red geometrical patterns, and the human 'arms' formed the handles, where the 'heads' formed the spouts. These pots are called 'anthropomorphic'.

All this was of great interest archaeologically, and in the usual way the majority of the finds had to be handed over to the Turkish government, though Mellaart was allowed to bring some sherds of the pottery back home with him. But from 1965 onwards examples of Hacilar pottery, including some anthropomorphic pots, began to appear on the antique market and were bought up by museums all over the world and by private collectors, for they are not only fascinating for their age, but they are rather charming, too. One of the museums which bought this pottery was the Ashmolean Museum at Oxford which purchased a remarkable 'double-headed' pot. But the Ashmolean was also the place where Mellaart had deposited some of the sherds he had officially been allowed to bring out of Turkey. Sinclair Hood of the Ashmolean Museum noticed certain differences between the pot the Museum had bought and the sherds Mellaart had given, and he asked Martin Aitken, the man who developed the thermoluminescence technique at the Research Laboratory for Archaeology and the History of Art in Oxford, to test the authenticity of the purchased piece against the known genuine sherds. There was no doubt that the pot showed clear differences from the well-marked thermoluminescence of the sherds, but Aitken felt, at that time, that he could not condemn unequivocally as fakes the Ashmolean pot, and three others known to have been bought at the same time.

There the situation had to rest for four years, but in those four years the antiquities markets in London, New York and various European centres were flooded with 'Hacilar pottery'. Some of these were obvious fakes of poor quality, others were believed to be genuine, but illegally excavated by local people. Some experts had reservations, however, and notable among those who were perturbed was Peter Ucko, the anthropologist, then at University College, London. He had started examining Hacilar ware from the stylistic point of view as early as 1962, and he was responsible for the withdrawal of a number of pieces from auction rooms, when he questioned their authenticity.

But by 1969 Aitken and his team were much more experienced with the techniques of thermoluminescence, and had examined hundreds of pottery samples for purely archaeological dating purposes. The reopened their enquiry into the Ashmolean samples, and this time were so convinced that the double-headed anthropomorphic pot purchased in 1965 was made in modern times that they decided to conduct as full a project as possible into many more of the pieces that had come onto the market, including specimens of Hacilar figurines. They got access to a total of sixty-eight Hacilar specimens, including samples from the British Museum, the New York Metropolitan Museum of Art, the Ankara Museum, the Israel Museum (to which they had been loaned by a Portuguese collector), many private collections in Britain and the U.S.A., and also to samples from the German antiquities market, provided by the Badisches Museum at Karlsruhe. The Ashmolean Museum also supplied further samples, but eventually the results on only sixty-six specimens could be published, because permission to publish was refused by two owners.

The conclusions were a great shock: only eighteen out of the sixty-six samples of Hacilar pottery were genuine. All the others showed by their lack of thermoluminescence that they had been fired in very recent times.

This baldly sums up a great deal of work, but we have examined in an earlier chapter the principles and qualifications of the thermoluminescence technique. The only major addition to this technique in this particular investigation was a special 'pre-dose' checking on many of the samples. This was done by separating bart of the clay sample and submitting it to a high dose of radiation pefore going through the normal measuring technique. The results from these sub-samples were used to check out the main sequence of results. The object of this operation was to cover against a 'defence' of the pots which seemed to be a possibility. There had for some years been stories that the Hacilar pots on the market were being discovered by local peasants who had found the site of the cemetery of the ancient village, but the graves in this cemetery were waterlogged by the natural spring which was believed to have been one of the attractions that originally induced men to settle on the site. These waterlogged pots were supposed to be completely 'plastic' when dug up, and the story was that the peasants were refiring them, probably in their bread-ovens, to make them firm and attractive for the black-market dealers. The story was probably concocted to account for the difference in the outside encrustations that was observable between 'new' pots and those dug up in Mellaart's properly recorded excavation, but such refiring might have reduced the thermoluminescence of the pots which would confuse the Oxford investigators. The pre-dose technique, however, showed that the pots declared 'modern' by the standard thermoluminescence measurements were in most cases undoubtedly modern. Aitken summarizes these additional results:

Consideration is given to the possibility that the Group II objects [Author: i.e. those found modern] are genuine ones that have been refired. However the most likely temperature to have been employed in the event of such treatment would not be sufficient to remove the thermo-luminescence. For 31 of the Group II objects the pre-dose technique gives evidence of recent manufacture that is valid even if the objects have been refired to a higher temperature. The remaining 17 do not exhibit the pre-dose phenomenon. Chemical analysis of 13 of these indicates that all but two are made from a different type of clay to that used for the sherds and figurines obtained by recorded excavation. It is concluded that out of the 66 objects, 48 are recent forgeries (7 bowls, 12 anthropomorphic vessels and 29 figurines).[1]

The chemical analysis that sewed up the last hole in the case made against the Hacilar pottery by thermoluminescence was done by Francis Schweizer using optical spectroscopy. Incidentally his report carried a possible implication that some of the Hacilar forgeries had been made with clay from that region though not from the same clay pits as the ancients had used.

The whole of this scientific work was confirmed by Peter Ucko on stylistic grounds. When it was published in the journal *Archaeometry* in July 1971, it made front-page news in the daily newspapers such as *The Times*, which had the headline 'Tests show forged Turkish prehistoric pottery in many world museums.'

This article drew attention to what archaeologists probably regard as most serious feature of the whole business. Archaeologists are probably quite happy to see the museums and private collectors getting their fingers burnt by finding they have paid thousands of pounds for worthless fakes, for they feel that by buying smuggled antiquities these collections simply encourage the looting and destruction of valuable sites. But Peter Ucko pointed out in his very last paragraph that the flood of material from Hacilar could have led to wrong historical conclusions if most of it had not been proved fake.

> . . . the risk that features of the unprovenanced material, if accepted uncritically, could affect our understanding of prehistoric Anatolian activity and beliefs. (See for example the 'Classic Mother Goddess' arm position with hands grasping the breasts . . .) Without these analyses, and the realization that many of the figures concerned are modern forgeries, an emphasis on the vulva might also be suspected for the Anatolian Neolithic.[2]

In other words, forgeries committed for gain on a large scale, could very well seriously mislead scholars into misinterpretations of the past.

Yet that same issue of *Archaeometry* contained the exposure of yet another major series of fakes involving the European and American art and antiquities markets during the same decade that the 'Hacilar' pottery had been deceiving people. This second series of fakes were 'Etruscan wall paintings' on terracotta and they, too, were exposed primarily by the techniques of thermoluminescence supported by stylistic typology and chemical analysis.

The history of Etruscan fakes is long and distinguished as such histories go. It goes from a sarcophagus 'manufactured' about 1860, through various pieces of sculpture and painting including famous examples such as the 'Warriors' produced about

1915, to a fabulous 'Diana' which was made about 1930. The reason for this long continuity of the forger's art is that archaeologists continue to find new Etruscan tombs similar to the magnificent discoveries of whole tomb-cities at Tarquinia, Cerveteri and Veii. It is known that there is every likelihood of continuing to find more tombs. The black market in Etruscan art can, therefore, always be expected to produce a string of new discoveries, some official and recorded, some illegally dug and possibly smuggled out of Italy (viz. the Greek 'krater' vase acquired by a New York Museum in February 1973), and a number of forgeries. Of those who illegally plunder the tombs they find, Professor H. Jucker, of the Archaeological Seminar at the University of Bern in Switzerland, says: 'The latters' [the plunderers'] activities have scarcely been deterred by any scruples of the Western art market, which seems to heed little the bitter experiences of the past century.' Professor Jucker was responsible for art-history and critical stylistic analysis which, together with the thermoluminescence technique, helped to expose the latest set of Etruscan forgeries.

These forgeries took the shape of four wall-paintings which appeared to have been stripped, terracotta and all, from Etruscan tombs. They came onto the market in the years from 1963 onwards. Technically they are called 'pinakes'. But although only four have been examined scientifically they are representative of many that appeared on the market, particularly in Basle, and in their publication exposing the fakes, the three authors take care to demonstrate the chemical composition of the terracottas, both those judged to be genuine and the spurious ones, so that further comparisons can be made.

Doubts about the many 'Etruscan' scenes began from an art-historical point of view, because many of them seemed to be essentially Greek in subject matter and style, as if copied from Greek originals. It was in 1967 that one in particular, the so-called 'Ambush of Troilus', became available for study. In 1969 the scientists of the Oxford Research Laboratory for Archaeology were called in, and a team led by Dr Stuart J. Fleming undertook to examine the terracotta pinakes by thermoluminescence. They got their samples by scraping a little of the plasterwork from the back of the slabs and pieces. They quickly showed that four of them were of recent manufacture, probably having been made within the previous twenty years at most.

The co-operation between the Oxford scientists and the Swiss art-historians was strengthened by getting J. Riederer of the Doerner Institute in Munich to examine the chemical composition of the paint. He used many techniques – emission spectroscopy, X-ray diffraction, infra-red spectroscopy and the ordinary microscope. His conclusion is brief but pointed: 'Pigment analysis revealed the imitators' inability to overcome the temptation to use painting media derived from modern sources. Such a failing is understandable in view of the vast technological knowledge that would be required to produce art in the truly original media.'

This story, too, was taken up by the newspapers. The *Sunday Times* of 1 August 1971 went a good way further than the original scholarly paper. The newspaper claimed that twenty-five such 'Etruscan' wall-paintings had been sold on the art market for about £10,000 apiece, and that there had been considerable pressure on the scientists to stay silent. Dr Stuart Fleming is quoted as saying, 'They demanded that we kept quiet because they had been told they would get their money back from the forgers if there was no publicity. Fortunately there were four who agreed that the whole thing should be exposed. They felt people should be warned.'

This may seem a long way from the academic discipline of archaeology; but archaeology and its finds have money value as well as scientific interest, and the science of the laboratory has shown it can check those who wish to use an archaeological disguise for fraud.

1. M. Aitken *et al.*, *Archaeometry*.
2. P. Ucko, Archaeometry, July 1971.

14

Crisis in the Profession

The image of the archaeologist has changed out of all recognition in the past twenty years. The popular picture, it is true, remains a picture of a man with a spade, or a man controlling gangs of 'natives' with spades; and every so often a hidden city with ruined temples and hoards of gold jewellery and wonderful carvings emerges from the desert sand or the tropical jungle. But the archaeologist's picture of himself has changed, and the self-image is the correct picture. True, some of the older generation of archaeologists still keep the old and popular view of themselves, half-ruefully, half-defiantly describing themselves as 'dirt' archaeologists, men who simply dig for their living.

But whether he digs or not, the modern archaeologist knows that he depends on the physical chemists in their 'radioactive'-labelled laboratories to provide him with dates from the radiocarbon clock or one of its successors and rivals. The present-day archaeologist may go into the field with a computer-terminal in his baggage or with a computer-programme guiding his plan of action. He may be going to a site which he cannot 'see' when he starts digging; it had been located by aerial survey or by proton-magnetometer readings. If he is in a large party the archaeologist may be accompanied by geologists, ecologists, plant-geneticists, botanists and zoologists. In his digging he will have to bear in mind the demands of the radiocarbon daters or the thermo-luminescence-daters. They will want chemical analyses or perhaps even samples of the earth and soil surrounding his finds. He may have to use the flotation process on almost every spadeful of soil he turns to find material for the palaeo-botanist. He will certainly have to bear in mind the work of anthropologists and ethnographers in his interpretations of what he finds. Ideally the archaeologist would be all these different sorts of scientist in himself. This is manifestly impossible, so he must either take some of them with him in a multi-disciplinary team, or he must know enough about their work and their working methods to provide them with data which is compatible with their different standards.

Gone are the days when a risky climb up a rocky cliff, some deft work with pencil or papier-mâché pressing, and a period of study back at home would enable him to announce the discovery of a forgotten civilization or the decipherment of a script. Gone, too, are the days when a careful record of the stratigraphy of a site, the careful measurement of the successive layers of ruins laid bare, and a general knowledge of the pottery types of northern Syria would allow an archaeologist to give a sufficiently comprehensive account of a city over three thousand years of prehistory, as Woolley was able to do in the Amq forty years ago.

There are, of course, many archaeologists, particularly in Britain and Europe, who bitterly resent this state of affairs, and Jacquetta Hawkes warned more than four years ago in an article in *Antiquity* entitled 'The Proper Study of Mankind' about the importance of 'Preventing the scientific and technological servant from usurping the throne of history'. Broadcasting in a B.B.C. series of programmes on the modern developments in archaeology in November 1971, she held to the position that this 'usurpation' was the chief danger:

That is my fundamental fear. Because, after all, it's what's happened in society at large. We are now all getting terrified of our way of life being more and more dominated by the machine and by the statistical outlook. And it seems to me sad, if this is also going to affect our understanding of the past. Because I've always seen pre-history and ancient history, one of their great purposes and justifications is they show the wonderful variety of achievement of man. All the splendid civilisations that there have been, always their completely different complexions, their different cultures. And I think at the present time there really is a tremendous need for more studies of prehistoric religions and the psychology behind them . . . The kind of argument you have over technical difficulties, disagreement as to whether carbon-dating is reliable in this way or that, and that type of argument tends – I think it's fair to say – to be rather more sterile than the historical type of argument in which you do gradually edge towards a kind of consensus of a possible historical truth.

Questioned about the possibility that the humanist point of view tended to recreate the past in the image of the present, Jacquetta Hawkes answered that, on the contrary, this was the failure of the scientific approach:

They want to see Stonehenge as a computer, and all our stone circles, which I believe to be primarily religious, to be mathematical machines of various kinds. I would say they were much more the ones who were using the past in a modern shape, forcing it into a modern shape. They vary, obviously very much, from extreme fanatics to people who are very close to me indeed, who really want to make the new scientific methods serve history, serve the writing of history. But some of the fanatics, I feel they are quite content to become skilful with their machines and they don't realize perhaps how difficult it is to serve two masters or two values at the same time. And I think there's a great danger of completely branching off into doing these technical investigations which have no relation to history at all. Scientists are rather optimistic; they always think they themselves and all society can have things both ways, and that tends not to be possible. Particularly, I would think, that quite a different type of young recruit will be attracted, therefore that there may be an almost complete disappearance of the humanist type of mind. I was very much encouraged to take this, if you like, reactionary view by the correspondence that followed that article of mine; because I had a lot of letters from students in different parts of the world saying 'Why didn't you write this before? We would never have decided to read archaeology at the University if we had known it was going to be all these statistics and analysis and so on.' So you see that that type of humanist thinking will be driven away from the subject and there might be a real danger that there will be no one great humanist who could weld it all together and make it into history.

Even those who are committed to the new methods see the dangers which Jacquetta Hawkes has outlined. In the same broadcast programme, Professor R. J. C. Atkinson, the expert on Stonehenge and Silbury Hill, one of the pioneers of the elucidation of prehistoric man's technology, admitted:

If one looks at these new methods simply as alternative approaches or, if you like, alternative tools, then it seems to me that one can get a great deal out of them as long as one always bears in mind that they are simply methods and they are not an end in themselves. A lot of these new methods of locational analysis of archaeological distributions, of classification by a numerical means, these are techniques which do involve some degree of numeracy. And, let's face it, the great majority of archaeologists are not as numerate as an educated person should be. And in consequence, at best they are very often unable to use these new methods in a critical way, and there is a danger of the method being used wrongly in the sense that it's applied to data which are not suitable for it. At worst, it can lead to what I would call gadgeteering, which is only in a sense another name for mistaking the means for the end, that you use these methods simply because this appears to be a more scientific way of approaching archaeology. It isn't of course. You can tell lies with numbers just as easily as you can with words.[1]

Whether we agree or not with Jacquetta Hawkes' point of view, she certainly picked out the two developments which are causing most of the problems in the archaeological profession at the moment: 'branching off' and 'education and recruitment'. Taking the second of these two problems first, it is quite clear to most modern archaeologists that the student, the archaeologist of the future, cannot possibly be taught the physical chemistry, zoology, botany, genetics, animal physiology, computer-programming and so on that will be needed in connection with his work. Yet the student cannot just become a trained excavator providing materials for all these other sciences. The problem was clearly set out by Professor G. W. Dimbleby, by training a botanist, in his Inaugural Lecture as the new Professor of Human Environment at the Institute of Archaeology in London University in 1965 – a position in which he succeeded Frederick Zeuner.

Archaeologists are always ready to admit how little they know about ecology, geology, pedology and all the other -ologies which are needed for a complete study of the environment. This is largely a legacy of the type of training which it has been customary for an archaeologist to receive, based on an arts or classics background and with no contact at any point with the sciences. It is difficult to see how it could have been otherwise in view of the motives which have driven us to take an interest in our early ancestors and their beliefs. Today, however, this lack of environmental understanding is at least widely acknowledged and regretted, and there are indeed some archaeologists who have acquired the appropriate background of knowledge.

The answer that Dimbleby developed is one that has become popular, even common, in universities and institutes on both sides of the Atlantic. On the occasion of his Inaugural Lecture Dimbleby said:

So far the development of archaeology has centred primarily upon man himself and his cultural achievements; its methods have been mainly comparative study of his artefacts – particularly of the relatively indestructible things he left behind: his earthworks, buildings, weapons and the utensils he used. In the methods of study and in the material studied, archaeology has been typically an arts subject. What, above all else, has made it so has been its focus upon man and his cultural development. If however, the point of the focus is changed, so that the object of study is the world in which man lived, and still lives, then the subject becomes a science comparable with geology, anthropology or ecology. After all, the research methods used in an arts subject, in so far as they involve observation, comparison and logical deduction, are also common to science . . .

Now almost every archaeological site will contain something more than the material for the study of human cultural development and the means of dating it. It may contain evidence about natural vegetation, soils, fauna and other features not necessarily of cultural significance (though of course they may be). I find it very disturbing to realize that every time a site is dug fundamental scientific information may be lost, simply because it is not looked for. I am not apportioning any blame for this; it is just unfortunate that archaeologists, by virtue of their background, are unaware of the potentialities, while scientists equally seem to have been unaware of the possibility that there could have been something preserved which might interest them. Today there is an increasing awareness on the part of archaeologists that significant non-archaeological material may be met with on their digs, but even when it is, the use made of it is so often inadequate. Charcoal, soil samples, samples for pollen analysis, all may be submitted to experts, but their reports are merely tacked on as appendices to the normal archaeological reports. Nowhere in the body of the report does the archaeologist discuss the significance of the special reports, even if he mentions them: the expert, for his part, contents himself with a brief factual statement, quite unrelated to the context. Somehow this cleavage must be removed, and it seems to me that the best hope is to train archaeologists who are capable of assessing the environmental as well as the archaeological data. . .

Archaeology, more than any other subject I can think of, with the possible exception of religion, stands to benefit from the integration of arts and science. These two unsatisfactory terms begin to lose themselves in this sort of context . . . and here much of the recent thought on the dichotomy between the disciplines – or cultures – shows to full advantage. Granted that specialisation has resulted in great progress in many directions, we are now in the position, not only in archaeology, where the centrifugal development is leaving a vacuum at the centre, so that no one person can easily perceive what is going on all round the periphery.[2]

Although Dimbleby sketched his answer to the problem when he first took the chair in 1965, it took a good many years more before it was possible to see that answer coming to a reality. It is hard and slow work to move or change the structure of any university, and it was not until 1972 that the first handful of students trained to study the human environment both past and present received their first degrees.

In America the situation, for rather different reasons, turned out to be remarkably similar. The Europeans who arrived in the fifteenth and sixteenth centuries found cultures which must be formally classified as Neolithic. Whether we refer to the Woodland Indians of North America or the high civilizations of the Aztecs and the Incas in Central and South America, the people were in the Late Stone Age or even earlier stages of development. Even when western, nineteenth-century, scientific/academic studies began in the U.S.A., and the ruins of the Central American civilizations began to be rediscovered, those Indian cultures which still remained alive on the continent remained essentially in that stage; this applies most notably to the Indian tribes of the southwestern U.S.A. American academics could therefore study living examples of what a European would call prehistoric man, and the first great drive was to study Indian languages and culture while they still remained extant. Anthropology was therefore the science, the discipline, which studied prehistoric, Stone Age, but contemporary man in America.

William A. Longacre, Associate Professor of Anthropology at the University of Arizona, would be described in Europe as an archaeologist, judging by the sort of work he does. In 1970, speaking to the American Anthropological Association, he said:

> Historically in America archaeology has always been considered a part of the larger discipline of anthropology. Its development is inseparably tied to the development of anthropology as a whole. The archaeologist deals with past cultures from an anthropological point of view and is concerned with making contributions toward the attainment of the goals of anthropology. As these goals have changed, so has archaeology changed.

In America, just as in Europe, though perhaps a decade earlier, the problem soon emerged of how to train the archaeologists of the future. Longacre continued,

> In fact there gradually developed a pattern of divergence in the interests of cultural anthropologists and archaeologists that was sometimes hard to reconcile. Archaeology students grew increasingly impatient with courses in the details of kinship and cross-cousin marriages, as they could see no relevance to their own interests in prehistory. Likewise, budding cultural anthropologists could see little utility in memorising sequences of past cultures and tool-types in archaeology courses. But the strong bonds of shared interest in the culture history of various parts of the world and the concern with the development of major cultural achievements such as agriculture and civilisation from a comparative point of view, continued.[3]

In the U.S.A. the practice of studying environment as part of archaeology, and then of including archaeology under the heading of 'environment', began mainly in the universities of the south-west. It is significant that one of Longacre's most important contributions to the development of archaeological thought in recent years – his contribution to the seminal volume *New Perspectives in Archeology* edited by the Binfords, which will be considered in greater detail in the next chapter – begins, immediately after the general introduction, with a major section entitled 'Socio-cultural Background and the Environmental Setting'.

The most common answer, then, to the problems of archaeology and the archaeologist facing the impact of the laboratory sciences and modern technology is to shift the focus of archaeology from the study of man and his cultures to the study of man in his environments. In relation to the environment, cultures are then seen as a factor in evolutionary situations. Longacre expressed it thus:

> Some archaeologists are moving away from a concept of culture that sees it as a set of norms or ideals which are internalised and shared. They argue that cultures are composed of highly interrelated sub-systems, and it is this highly systemic whole that stands between the biological population and the total environment. As such, culture is not shared, but participated in differentially, and its various subsystems are subjected to selective pressures that lead to adjustments and changes as well as stability. Thus the focus is upon processes of cultural stability and change, and the goal is to better understand the nature of culture through comparative analysis.[4]

This seems to offer a long-term compromise solution between the two camps – the scientific and the humanist – and to offer something for all types of student.

But the experts in separate disciplines will still be needed to help the archaeologist with specific problems, and here we come to Jacquetta Hawkes's problem of 'branching off'. The difficulties come after the first honeymoon period between archaeology and each new expertise that archaeology calls to its aid. The expert on pollen identification is at first delighted to find that he can make a positive contribution to archaeology. He is excited by this new use for his professional expertise; he co-operates willingly and devotes many hours of his time to the new adventure; he is thrilled to go out with the archaeologist to the fresh air and remote situation of the dig. He will even put up with the damp or the heat or the mosquitoes or whatever are the particular disadvantages of the site. But after this stage of excitement has been passed the problems begin. Many other archaeologists realize the value of the new technique and the pollen-analysis expert is flooded with work, so that he can longer get on with his own job. Then it becomes the case that no archaeological investigation is complete without a pollen analysis, even though the archaeologist is not really interested – in this particular investigation – in either the pollen or the problems to which pollen might give the answers. So the pollen analysis is tacked on to the main report as a mere appendix. The pollen analyst realizes that the work he has done, perhaps rather grudgingly, is not even essential; and why should he toil away at laborious countings and microscopic peerings when he has work of his own to do? That is putting it at its worst. The more interesting outcome is when the 'outside expert' finds that his first co-operation with archaeology has brought out a whole series of fascinating questions he can ask in his own original field. His first analyses of pollen counts from sites in north European bogs, which he performed as a botanist, have suggested to him that he has found a technique for examining the pattern and timing of the spread of the elm tree as the hardwood forest moved northward behind the retreating ice. It is a botanist's problem, which can be solved by botanical techniques, as long as the archaeologist will provide the data from his diggings. This is 'branching off'.

It is a problem not confined to archaeology. The technique of tissue-typing in the field of immunology is a good example from another field. The principles of tissue-typing came from cancer research, when it was discovered that some individuals are more alike to each other, in an immunological sense, then they are to the rest of mankind. In practice this means that the like individuals do not reject each others' cells and tissues as violently, or as quickly, as the unlike individuals. Furthermore this likeness and unlikeness can to some extent be examined, measured, and accounted for, in terms of certain factors in or on the cells. These discoveries were of the greatest importance to the surgeons and clinicians who, some five years ago, were beginning to practise transplant surgery on a large scale, and it is nowadays regular procedure to 'tissue-type' those patients who need transplants, so as to provide them with organs from donors whose 'tissue-type' is as like as possible to their own. For the research workers who had discovered the phenomenon of tissue-types, and those who had joined in the development of the details, what at first had been a great and exciting adventure in collaboration with the surgeons, became a threat to their professional lives. Their laboratories virtually ceased to produce research results; they became factories, or production lines, turning out tissue-type 'profiles' of prospective recipients and donated organs, often at high speed and under great pressure; while, all the time, the research men themselves wanted, instead, to use their new technique of tissue-typing to investigate other problems of the human body. So the transplant surgeons have seen their expert tissue-typers 'branching off' into other work which has no connection with surgery or even with present-day clinical practice.

So it is with the archaeologist. He sees the plant-geneticist, whom he has called into help with the problem of the origins of domesticated wheat, becoming more interested in the botanical and evolutionary problems of the development of plants under the selective pressures of domestication. Perhaps the expert even demands more samples from the archaeologist for the work on the expert's own research, than the archaeologist cares to provide. Indeed this 'branching off' process has become so marked over the crucial questions of the origins of agriculture that the situation has now developed into one where the archaeologist is regarded by biological and botanical scientists as the key figure in *their* hunt for the scientific facts about the origins of the important crop plants of mankind.

This swaying of the balance between the originating discipline and the 'branching off', or temporarily subsidiary discipline, can be well illustrated by the work of Professor Dimbleby. He started life as a botanist and was given the research task of finding out why it was so difficult to grow trees on certain moorlands and heathlands in Britain – these stretches of country seemed almost lethal to trees. He soon established that the problem lay mostly in the nature of the soil, which was called a 'podzol' – a soil in which all the iron and mineral nutrients had been washed out of the top layers, leaving an acid, ashy-grey, mixture above an extremely tough, almost solid, layer of minerals, a 'pan'. The standard belief, as taught in the schools and universities in those days, was that such a soil was formed by British (and north European) climatic conditions acting where there was not enough chalky, calcareous mixture in the soil. This implied that it was inevitable, and nothing much could be done to cure it. But then why, the young Dimbleby wondered, were there so many earthworks and barrows in those areas, which had presumably been barren and infertile ever since the British climate first started acting upon them?

So the botanist perpetrated an archaeological sin; he dug a pit right down through a round barrow until he came to the original soil on which the structure of the barrow had been piled. The soil buried under the barrow showed no trace of the leaching process – the washing down of iron and so on – which it should have done if it, like the soil all around, had been developing into a podzol by natural processes ever since the Ice Age. Furthermore the pollen in this 'preserved' soil showed that when the barrow had been built the land had not been heathland at all, but had been oak-woodland with heathy clearings. Dimbleby said of this discovery:

Using this method it has since been shown that soil deterioration following forest clearance and a period of primitive cultivation is very widespread; our poor soils of today are often the direct result of human influence starting in prehistoric times. Furthermore we know now, that if left alone the vegetation and soil will gradually return towards the more fertile condition which previously existed. But the real point of this story lies in the fact that, unbeknown to me, archaeologists already had the relevant information: van Giffen had found, time and time again, that on the Dutch heaths there were unleached soils under Neolithic barrows, but podzolised soil under Bronze Age barrows. Indeed he used this as a criterion for judging their age. As far as I know he did not draw the ecological conclusion from this; certainly the ecologists and pedologists had not assimilated the facts that he had unearthed. Had they done so it would have made a big difference to what was being taught as the fundamental principles of soil genesis, and I, working in an applied field, would have been saved a great deal of time in working out for myself the principles which I should have been applying.[5]

Later Dimbleby turned his discovery, made in the course of applied botany and forestry work, more directly to archaeological purposes. Studying a site on Great Ayton Moor in north-east Yorkshire, where there are Neolithic, Bronze Age and Iron Age barrows, he made pollen analyses of the buried soils under the differently-aged barrows. These showed clearly that the area had been woodland when the Neolithic men built the earliest barrows; gradually more and more of the land come under cultivation through the Bronze Age and cultivation reached a peak in Iron Age times as shown by the proportion of cereal pollen. But at that same time the heather pollen first appeared in quantity, and it increased, as a proportion of the whole, under later Iron Age earthworks. Today the moorland, though it is one of the loveliest areas in Britain, is treeless and heather-covered. Dimbleby comments, 'Such information about a specific site has never as far as I know been achieved by conventional ecological methods. The method of approach, in fact, is parallel to the search for the origins of maize.'

Confirmation of the general idea that much of Britain was stripped of its woodland cover and rendered infertile by the ever more intensive farming of Bronze Age and Iron Age has come only a few weeks before writing of this chapter (February 1973), and from a most unexpected quarter. A study of the frequency in the soil of the remains of a beetle which favours the wood of oak and beech forests shows that this little creature has become far less common in the British habitat since Bronze Age man lived here.

The question remains, however, whether botany, pedology,

archaeology or prehistory benefited most from the discoveries of buried soils under the north Yorkshire barrows. Depending on the answer we give, we will form corresponding views about the requisite training for our future archaeologists. Indeed there is even a suggestion that archaeological work may have utility, as well as increasing knowledge in general. Dimbleby concluded, 'I am not suggesting that studies of the past conditions will solve present problems, but I do believe that, taken with what we know and are still learning about ecological processes, they may tell us where things have gone wrong. Until we know this, efforts at restitution will be of a hit-or-miss nature and much expenditure of money and effort may be stultified.'

But as well as the lessons to be learned from archaeology about the history of ecological systems, there is an even more pressing development which is dragging the archaeologist into direct contact with the problems of the modern world. This again is changing and increasing the demands on the archaeologist. The problem is that of the 'rescue' dig.

The redevelopment of cities in both Europe and America and the rapid extension of motorways, highways, thruways, autostradas and pipelines through great swathes of undeveloped countryside and farmland are continually turning up, or threatening, archaeologically valuable sites. Sometimes the civil engineers actually reveal previously unsuspected remains; a new motorway through Gloucestershire exposed dozens of small Roman farmsteads in an area previously thought to have been virtually unoccupied by the Romanized Britons. Yet when a city block is to be redeveloped this may be the one chance, for several generations, to find out what remains of previous buildings lie below.

The problem of city redevelopment might be thought to be confined mostly to Europe, and it is certainly most pressing there, but in Central and South America there is a tremendous amount of archaeologically valuable material underneath modern cities – Mexico City itself is the most obvious example. Even in North America, with its comparatively short history in city terms, there is often much that can be recovered about the history of the early settlements when the bulldozers and mechanical shovels get to work along the river banks. It is often in the largely unexplored (archaeologically) countryside of North America that some new engineering project reveals and/or threatens some previously unknown remains of Indian cultures.

Whether in city or countryside the archaeologist is either called to the 'rescue' or has to fight to perform the rescue – at least to recover what information he can before the site is sealed for generations under tons of concrete. The archaeologist then must become politician, at least on the local scale. He may have to drum up local votes and political support to pressurize the local authority, or to persuade the developer, at least to grant a stay in the starting date, so that excavation and archaeological study and recording may be done. Funds must be raised to finance the digging, or to pay for the preservation of the remains; volunteers must be raised to help in the work. Very often, too, the archaeologist must be, to some extent at least, an engineer, to perform his work in the deep basement diggings of some city block or among the concrete and cuttings of a new highway.

Surprisingly, and contrary to experience in many other areas, it is getting easier to find money for rescue digs. The basic cause of this is undoubtedly the growing understanding, at least in most western countries, of the importance of preserving our various heritages. On the one hand this makes the raising of voluntary funds for rescue digs easier. But perhaps even more important it has changed the political atmosphere and made it possible to effect new legislation and new action by central or local governments. Thus in Sweden, in some states of the U.S.A. and in several of the West German *Länder* there is legislation making it either compulsory or at least possible to include clauses in development contracts by which the developer must pay all or part of the costs of archaeological work necessitated by his operations. In some cases a fixed percentage of the contract price is allocated to archaeological purposes. In other countries there is no such legislation; France is an example where the archaeological rescue situation is 'chaotic', in the word of one expert. In Britain there is consideration of legislation which, while not forcing the developer to provide money for rescue operations, would at least prevent him from forbidding rescue. At the moment a developer can legally forbid archaeologists from having access to the site. The legislation envisaged would set up laws of access and give the potential rescuers some rights in getting time for their work. It has to be remembered that, however desirable rescue or preservation may be, a three- or four-week delay in getting men and machines to work on a big redevelopment or on a motorway can cost the developer or contractor many thousands of pounds. With better access, or more time provided, the archaeologist will in future have many more agonizing decisions to make, for the greater opportunities will thrust on him the burden of deciding whether recording the find will be enough for posterity, or whether he must make even greater efforts and fight for preservation. He will have to get his priorities right.

The size of the rescue problem can be judged by the fact that in the present year (1973) the British government is spending £200,000 and local authorities are spending an equal amount. In 1974, it is expected that this sum will top the £1 million mark from the central government alone. This doubling of the money available will make it possible to set up a national organization to co-ordinate rescue archaeology all round the country, and this will reinforce the present non-system of local *ad hoc* committees being set up in each area, or even for each site, when a need for rescue suddenly arises.

The new national rescue service will probably take the form of some fourteen regional centres, each with a small number of fulltime professional archaeologists, and each carrying some of the equipment required. The necessity for a regional system is that the scientific equipment demanded by modern archaeology is so expensive that it cannot be provided for every single local area. It is admitted in government circles that the public support for rescue operations and the political lobbying by archaeologists and others has been effective in getting the scheme set up. The archaeologists, as a profession, would have preferred a State Archaeological Service (such services can be found in East

European countries as well as in the most far-sighted western European nations). This proposal was discussed at a number of meetings but stood little chance of being accepted in the climate of opinion in Britain, where the political parties are always embarrassed by publicity about any increases in the number of civil servants.

The rescue archaeologist himself is usually seen wearing a civil engineer's safety helmet. He has a small and enthusiastic band of volunteers with him – housewives, students and local school-teachers mostly, some of whom may well have quite a bit of experience as excavators. The archaeologist may well have borrowed or hired a small mechanical digger or trenching machine to shift the bulk of earth and debris, and it hardly looks out of place next to the bigger machines of the civil engineering contractors waiting to get on to the site. The best the rescue archaeologist can hope for is usually to get a fairly full recording of the main levels of occupation and the chief structures underneath the new development. If he is lucky he will get some small finds of artefacts or coins which are worth preserving.

But bigger things can be achieved. A notable example of a recent triumph is at Dover, where a hurried dig in front of the progress of a new urban motorway revealed the hitherto unsuspected fleet-headquarters of the Roman Navy in the English Channel. This led to reconsideration of the development proposals and agreement has now been reached to raise the whole level of the new developments by several feet so as to preserve the Roman and medieval remains that have been found, in this historic 'port of entry' town.

In Central America rescue takes on a different guise. There it is mostly a desperate attempt to ward off robbers and smugglers before they break up the site and every artefact on it to sell to unscrupulous private collectors the world over, since pre-Columbian art has become an expensive fashion. There is so much material hidden in the jungle that the best the archaeologist can usually expect is to be able to record the broad outlines of the site and the most obvious stelae and sculptures before they are carried off.

In such desperate situations, faced by the philistines of our modern age the rescue archaeologist may well envy his more academic colleague who is recording a grid of readings from the caesium-magnetometer, and feeding the results of his day's work into the computer-terminal on the site. But both men (or women just as likely) are so far removed from the 'art-critic' archaeologist placing pottery types in a chronological sequence to match the twenty-five habitation levels of a Middle Eastern 'tell' that it is quite justifiable to say that there has been a revolution in archaeology and among archaeologists.

1. The quotations by Jacquetta Hawkes and Professor R. J. C. Atkinson are taken from their broadcasts in the B.B.C. series 'Rewriting Man's Prehistory', November and December 1971.
2. Reproduced by permission of Professor G. W. Dimbleby.
3. W. Longacre, 'Current Thinking in American Archeology'.
4. W. Longacre in *New Perspectives in Archeology*.
5. Dimbleby, Inaugural Lecture, Institute of Archaeology, London University, 1965.

15

Crisis in the Theory

While the nature of the archaeologist is being questioned on the one hand, the nature of archaeology is also being questioned – and mostly by archaeologists. Once again the process began rather earlier in the U.S.A. than in Europe. The reasons for the dissatisfaction with the current state of affairs were rather different on the two sides of the Atlantic, but the conclusions were broadly the same, and the result, too, has been the same. This result is widely called 'the new archaeology'.

There are many archaeologists, especially among the senior members of the profession, who deny that the new archaeology is really new; every generation believes its methods and ways of thinking are radically new, they claim. But the chief advocates of the new archaeology are those who also insist on its newness. They insist it represents a true break with the past. They are so convinced of this that it is most useful to take them at their word.

Their words are strong words. Colin Renfrew, one of the apostles of the 'new archaeology' in Europe, Professor of Archaeology at Southampton University, opens his latest book with the following barrage:

> The study of prehistory today is in a state of crisis. Archaeologists all over the world have realized that much of prehistory, as written in the existing textbooks, is inadequate: some of it quite simply wrong. A few errors of course were to be expected since the discovery of new material through archaeological excavation inevitably leads to new conclusions. But what has come as a considerable shock, a development hardly foreseeable just a few years ago, is that prehistory as we have learnt it is based upon several assumptions which can no longer be accepted as valid . . .

It has been suggested indeed that the changes now at work in prehistory herald the shift to a 'new paradigm', an entire new framework of thought, made necessary by the collapse of the 'first paradigm', the existing framework in which prehistorians have grown accustomed to work.[1]

In Europe there has indeed been a shock. In America the 'new archaeology' began its arrival more slowly, and rather sooner. Let us consider this steadier development first.

The aims of archaeology have developed through the years. The first stated aim was the reconstruction of the 'cultures' of the past. Then it was realized that there was a danger of 'losing the Indian in the artefact' and the reconstruction of the life-ways, the actual modes of living, of extinct people was added as a second great aim of archaeology. In the 1920s and 1930s a third aim was added: the archaeologist should study and explain the changes and developments in cultural history, it was felt, and he should find motives, causes and factors behind the changes so that from his work and the work of contemporary sociology there would emerge the laws of cultural dynamics – the hopes and desires of both Marxists and anti-Marxists can be seen behind this new aim for archaeology.

But in the late 1950s and early 1960s a severe wave of pessimism swept across the archaeological world concerning the progress that was being made in achieving these aims. 'There began to appear in the literature a general dampening of enthusiasm of those who some twenty years earlier had called for the archaeologist to turn his attention to processual investigations. There was a similar pessimism expressed in the writings of British scholars . . .', writes Lewis Binford. He quotes the Presidential Address of Gordon R. Willey to the 1962 meeting of the American Anthropological Association, in which Willey devoted much of his speeeh to commenting on the lack of progress in gaining understanding of the processes of cultural history and the changes therein: 'Certainly the answers to the causal questions as to why the ancient American civilisations began and flourished as they did and when they did still elude us, and what I can offer . . . will do little more . . . than describe and compare certain situations and series of events.'[2]

One of the great stumbling blocks that archaeologists came across in attempting to reconstruct the cultures, life-ways and cultural changes of prehistoric peoples was the awareness that it was thoroughly unsatisfactory and unscientific to extrapolate backwards from the ethnographic data obtained by studying extant primitive peoples and from this to draw conclusions about extinct peoples. Typical of the dissatisfaction of these times is a passage written in 1963 by Don Brothwell and Eric Higgs:

> It is also a picture drawn from our knowledge of primitive peoples being in recent times in 'cultural slums' and isolated by circumstance. Nor is it possible to believe easily that when artefacts dispersed they did so because a people migrated. But by such a hypothesis, which is an attempt to trace tradition in time and space, one gets a prehistoric picture of invasions and mass movements of peoples, and wars, and by a pyramid of hypotheses all the trappings of the historical picture. The tail of history wags the prehistoric dog. This hypothesis has resulted in considerations of the most devious and subtle intellectual complexity. It is a hypothesis which can still be made. But more can come out of archaeology than a faded history. It is not in accord with, nor does it satisfy the climate of thought of the present time, which asks not 'what happened in history?' but 'why?'

Of course these were precisely the years in which the value of the other scientific disciplines in archaeology were beginning to be felt. They were precisely the years in which radiocarbon dating really began to pour forth enough absolute dates to make the older methods of setting up relative chronologies seem often irrelevant; they were exactly the years in which MacNeish and his colleagues in Tehuacan were showing what an interdisciplinary approach could do. Yet the impact of these other sciences was very hard for the organization of traditional archaeology to cope with. Brothwell and Higgs have this to say:

> But there has been a further serious problem affecting scientific research in archaeology. There was a tendency for each discipline, having snatched off its special research piece, to consider it in the light of its

own core concept. Its own core concept was based on the consideration of its data in isolation, a hypothesis as to what tended to happen and what might happen if by good fortune its own material had been left to a more or less uninterrupted course unaffected by other factors in the biotope such as man. A hardened belief in restrictive practices, the worst kind of professional trade-unionism, grew up and became an article of faith. Only a rash and perhaps untrustworthy scientist or archaeologist would dare to cross a disciplinary territorial boundary, for if he did he could expect it to be defended fiercely and in the ensuing contest he would have little chance of survival. This is an example of true animal behaviour in the defence of territorial rights which it is as well to remember. The result of all this was that the data from archaeological sites became fragmented away into a variety of separate enquiries which overlapped occasionally and fortuitously at the peripheries. A horrid porridge with little meaning was created of anthropocentric, phytocentric and zoocentric disciplines, and indeed who can blame the few who have resisted such an approach if they declare that there is indeed no meaning to it?[3]

In America the interdisciplinary problems were dealt with more easily and more happily, almost certainly because, there, archaeologists considered themselves to be primarily anthropologists, and had not descended by tradition from people whose primary function had been the collection of museum exhibits. But correspondingly, just because they were more closely related to 'natural scientists', American archaeologists became aware more rapidly of the weaknesses of the inductive system. It has been traditional to believe that scientific processes are based on induction, but induction is logically unsound, as was proved in the seventeenth century by Locke. Like scientists in other disiplines, American archaeologists became more interested in the logically tighter deductive systems.

And it was in America that the theory of the new archaeology was first developed. In fact Braidwood, the great American excavator of Middle Eastern sites such as Jarmo, had used a method of 'hypothesis-testing' in his search for the origins of agriculture in the Fertile Crescent. The new archaeology originated in the Middle West, round Chicago, but it has now become centred on the south-western universities of the U.S.A., where the big multi-disciplinary 'environmental' schools of archaeology have been built up.

The essence of the new archaeology is that it sets up hypotheses and at the same time demands the setting up of criteria by which these hypotheses can be tested. Binford writes:

We can infinitely expand our knowledge of the lifeways of living peoples, yet we cannot reconstruct the lifeways of extinct peoples unless we employ a more sophisticated methodology. Fitting archaeological remains into ethnographically known patterns of life adds nothing to our knowledge of the past. In fact such a procedure denies to archaeology the possibility of dealing with forms of cultural adaptation outside the range of variation known ethnographically. In view of the high probability that cultural forms existed in the past for which we have no ethnographic examples, reconstruction of the lifeways of such sociocultural systems demands the rigorous testing of deductively drawn hypotheses against independent sets of data. This perspective is in marked contrast to the epistemological basis of traditional method . . . So long as we insist that our knowledge of the past is limited by our knowledge of the present we are pointing ourselves into a methodological corner. The archaeologist must make use of his data as documents of past conditions, proceed to formulate propositions about the past, and devise means for testing them against archaeological remains. It is the testing of hypotheses that makes our knowledge of the past more certain, and this is admittedly a difficult business. Archaeology as part of anthropology, and anthropology as a social science are often guilty of the charges made against them by the 'harder' scientists: 'The most important feature about a hypothesis is that it is a mere trial idea . . . until it has been tested it should not be confused with a law. . . The difficulty of testing hypotheses in social science has led to an abbreviation of the scientific method in which this step is simply omitted. Plausible hypotheses are merely set down as facts without further ado' (Wilson, E. Bright Jr., *An introduction to scientific research*, McGraw-Hill, New York, 1952, pp. 26–7)'. Traditional archaeological methodology has not developed this final link in scientific procedure. For this reason, reconstruction of lifeways has remained an art which could be evaluated only by judging the competence and honesty of the person offering the reconstruction.

This quotation is taken from *New Perspectives in Archeology*, edited by Lewis and Sally Binford, which has proved to be one of the most important modern works in archaeology. It is certainly the first full-scale and formal statement of the position and objectives of the 'new archaeology'. In fact the contents of the book mostly consist of papers presented at a one-day symposium which was part of the 1965 meeting of the American Anthropological Association. The idea of this presentation of the work and thinking of the people who were coming together out of the depression of the 1950s and early 1960s with a set of new and rather similar ideas was the brainchild of Lewis Binford, the greatest exponent of the new archaeology, Sally Binford, Stuart Struever and William Longacre. The ferment was there, and this presentation and the book that grew from it crystallized the new movement and set out its theories publicly for the rest of the archaeological profession to digest. Much of Binford's introductory chapter therefore expresses a general point of view shared by most, if not all, of those who were with him in proposing a 'new archaeology':

> We assert that our knowledge of the past is more than a projection of our ethnographic understanding. The accuracy of our knowledge of the past can be measured; it is this assertion which most sharply differentiates the new perspective from the more traditional approaches. The yardstick of measurement is the degree to which propositions about the past can be confirmed or refuted through hypothesis testing – not by passing judgment on the personal qualifications of the person putting forth the propositions. The role of ethnographic training for archaeologists, the use of analogy, and the use of imagination or conjecture are all fully acknowledged. However, once a proposition has been advanced – no matter by what means it was reached – the next task is to deduce a series of testable hypotheses which, if verified against independent empirical data, would tend to verify the proposition.

This manifesto of the new archaeology concludes:

> Many of the authors in this volume would agree that advances in achieving the aims of archaeology necessitate the enforced obsolescence of much of traditional theory and method, and thus many of the papers in this book are radical in the original sense of the word. If we are

successful many traditional archaeological problems will prove to be irrelevant and we will see an expansion of the scope of our question-asking which today would make us giddy to contemplate. Despite a recent statement that one should not speak of a new archaeology since this alienates it from the old, we feel that archaeology in the 1960s is at a major point of evolutionary change. Evolution always builds on what went before, but it always involves basic structural changes.[4]

The new archaeology is obviously an attempt to provide an archaeological theory – a methodology for the discipline – which will make it, as a discipline, independent. It is therefore, in a sense, an answer to Jacquetta Hawkes's plea that archaeology should not become just the servant of the experts' techniques. Yet plainly the new archaeology will make archaeology again into a science and take it away from its traditional arts bias, at least in the European sense.

The new archaeology has come to Europe, however, largely because of a complete collapse of the traditional outline of European prehistory – the outline that was provided by the traditional 'arts' approach to the subject. Evidence provided by the newer scientific techniques, and notably by radiocarbon dating, has shattered the standard view of what happened in the five thousand years before the Christian era. What view will replace the old outline no one yet knows. Everything seems to be in the melting-pot, and fresh evidence keeps cropping up which throws things into still further confusion. We are in the middle of an exciting revolution.

Up to 1970 the whole of European prehistory had been considered by almost every scholar in the field as 'the irradiation of European barbarism by Oriental civilization', as V. Gordon Childe said. Childe was undoubtedly the most influential prehistorian of his, or perhaps any other time, and we have seen the importance of his concept of the Neolithic Revolution. His theory of the origins of European civilization has been called 'modified diffusionism' in that it rejected the fantastic ultradiffusionist ideas of people like Sir Grafton Smith, who held that all civilizations in all parts of the world could be traced back to Egypt, and likewise it rejected the nationalistic and chauvinist views of the German

Gustav Kossina, and others who, in the interests of political ideologies, tried to find the origins of civilization in their own countries. Diffusionism also replaced the rather stark 'evolutionism' of the late-nineteenth-century authors.

For almost as long as archaeology has been practised in Europe it was felt that most of the important steps forward had come about as the result of the influence of the great and ancient civilizations of the Middle East. Childe, postulating the Neolithic Revolution, and believing with good reason that this had occurred in the Fertile Crescent, gave sound reasons for civilization having begun in the Middle East, and, building on that premise, proposed two sensible and rational 'routes' by which the diffusion of agriculture, metalworking techniques, and building techniques could be traced westwards and north-westwards from the Middle East across Europe. These routes would also provide a chronology for European cultures which could be based on the well-established chronologies of the Middle East and the Mediterranean civilizations of Greece and Crete, which were based on connections with written Egyptian history.

One of these lines of argument, and diffusion, was provided by the megalithic tombs of western Europe and the Aegean; we have noted Glyn Daniel's discussion of these monuments, and he, broadly speaking, accepted the Childe hypothesis. The idea grew that the Spanish megalithic tombs developed from the Bronze Age tombs of Crete, which are very similar. The Cretan tombs were dated to about 2500 B.C. and so the Spanish tombs were a little later. The suggestion was that Aegean colonists had spread out to Spain, taking their architectural styles with them, and taking, too, their religion in the shape of the 'Mother Goddess' symbols, and their skill in metalwork which was shown in the development of the Iberian metal industry in the years from 2500 B.C. onwards. The great 'tholos' tombs of Mycenean Greece were considered a separate development, perhaps from the same origin, and were dated, firmly, to about 1500 B.C. through the connection of the Mycenean civilization to Crete, and the dating of Cretan remains by firm contact with Egypt. The Maltese 'temples' which also figured in the story were dated to about 1500 B.C. because sculpture with handsome spiral motifs in Malta resembled similar work in Crete and Greece which was dated to between 1800 and 1600 B.C. The Megalithic work and tombs in France and Britain and Denmark were all assumed to have developed in a steady succession as the influence or even the colonists themselves spread north from Spain. So these tombs were dated from 2500 B.C. to as late as 1800 B.C. Stonehenge, by similar reasoning, was dated as having been built between 1900 B.C. and 1500 B.C. The only crack in this argument was that Glyn Daniel allowed that there had been a separate and primitive beginning in the erection of large stones and graves in Denmark, which had met the incoming new ideas as they spread round the coast of Europe.

Childe's second route went across Turkey, through the Balkans to the Danube. It was based on the remains of the Vinca culture in Yugoslavia where Neolithic artefacts of various sorts seemed to be connected with one of the early Bronze Age levels at Troy, a level which had been dated to around 2700 B.C. There were also clay sculptures at Vinca and other 'artistic' remains which seemed to be similar to Bronze Age products in the Aegean. There was therefore a picture of the spread of farming and metalworking techniques from the Middle East into central Europe up the Danube line, and the extensive archaeological finds of the Danubian culture were all taken to fit into this picture of the first Neolithic farmers spreading in a slow wave northwards and westwards.

Childe himself well realized, and made clear in his writings, that there were two basic assumptions underlying his framework, and that both these assumptions could be challenged, at least theoretically. The first assumption was that if developments on the same lines at about the same time in two different regions appeared to be parallel, then they were not independent innovations in both regions. They must have had some common ancestor from which the technique – be it of metalworking or megalithic tomb-building – or, at least, the idea of the technique had diffused. Secondly he assumed that all diffusion in Europe had been from east to west. Since the only method of dating that could be applied to anything in prehistoric Europe was to relate it to the dated civilizations of the Middle East, this really meant that the only reconstruction that was rational had to accept both his assumptions.

Radiocarbon dating, by providing dating independent of connections with Middle Eastern civilization, provided the first reasonable way of checking Childe's assumptions. The first battery of radiocarbon dates merely confirmed Childe's general hypothesis. It is true that these first radiocarbon dates caused the 'revolution in the Neolithic Revolution' by placing the origins of farming in the Middle East back to 8000 or 10,000 B.C. where previously they had been 'guesstimated' at 4000 or 4500 B.C. But as far as Europe was concerned this simply gave more time for diffusion of eastern civilization to take place. The first radiocarbon dates for the megalithic tombs of western Europe gave their time of construction as about 2500 B.C., and though this was a little earlier than expected it did nothing to shake the structure of European prehistory. Just two minor snags cropped up. First, radiocarbon suggested that the megalithic tombs in France, and particularly in Brittany, were earlier than the other megalithic tombs to the south and to the north. Indeed dates of some tombs in Brittany seemed to be about 3000 B.C. However, 'most scholars simply assumed that the French laboratories producing these dates were no better than they ought to be, and that the anomaly would probably disappear when more dates were available', as Colin Renfrew wrote.

Even more surprising were the dates from sites relating to the Yugoslavian Vinca culture, which had provided one of Childe's most important linkages between the Middle East, the Aegean and central Europe via the Danube. These gave Vinca a date of about 4000 B.C., and therefore implied that the copperworking and clay statuettes of the Balkans were made 1,500 years before their supposed ancestors in the Aegean. This was so ridiculous from the prehistorian's point of view that some prehistorians, led by the Yugoslav Milojcic, claimed that the entire carbon-dating system must be wrong.

These European dates were rather small beer in the world picture that was interesting the physical chemists who had invented radiocarbon dating. What worried Libby and his collaborators were the Egyptian dates. There could be little doubt about the dates given to Egyptian objects based on traditional methods – there might be inaccuracies of a few decades but no more. Radiocarbon consistently gave dates for Egyptian objects that were too young; that is to say an Egyptian object of 2000 B.C. might be given a radiocarbon date of 1800 B.C., and further back the differences became more alarming. Libby himself argued at first 'that the Egyptian historical dates beyond 4,000 years ago maybe somewhat too old, perhaps five centuries too old at 5,000 years ago . . .'. He pointed out that the first astronomical fix by which the accuracy of Egyptian dating can be checked by our own modern time was at about 4,000 years ago, i.e. 2000 B.C. approximately.

The physicists were wrong at this point and the traditional historians were right. We have seen in Chapter 4 how the 1969 Radiocarbon Dating Nobel Symposium produced a consensus on recalibrating the radiocarbon dates by the tree-ring dating system. The really important thing for the prehistorian of Europe is that the traditional historic dates for Egypt are now accepted as correct, and the original radiocarbon dates for Egyptian objects were wrong. But if we recalibrate the radiocarbon dates for Egyptian materials using the tree-ring calibration then the radiocarbon dates and the historical dates come right into line. Furthermore the radiocarbon dates for Cretan and Aegean objects, which had been given too young by radiocarbon at the first attempt, also now line up with the historical dates, and the whole carefully built chronology of the Middle East, Egypt and the east Mediterranean is fully confirmed by the recalibrated radiocarbon dates.

The corollary of this is that all the radiocarbon dates for the Balkans and western Europe, which had at first seemed to fit so well into Childe's theory, must now be seen to be wrong by several centuries. The whole diffusionist theory of European civilization coming from the east is thrown down. The great and glorious Spanish megalithic tomb of Los Millares which was originally radiocarbon dated to 2350 B.C. must now be put at 2900 B.C., long before its supposed predecessors in Crete. Virtually all the megalithic tombs of north-west Europe must have been built before 2500 B.C. and the tombs of Brittany must be put back right into the fourth millennium. The final stage of Stonehenge, previously supposed to have been 'masterminded' by Mycenean craftsmen from Greece about 1500 B.C., must have been built several centuries earlier, long before there was a Mycenean civilization in the Greek Bronze Age – for Greek dates, tied into the Egyptian calendar, remain the same, unaffected by the radiocarbon recalibration.

In the Balkans the situation is even more revolutionary. The Vinca culture must now be dated to 4500 B.C., and therefore there was metalworking in the Balkans and the Danubian area before it had become widespread in the Aegean. This date for Balkan metalwork is still later than the start of metalwork in the Middle East – but even if we can find some other route by which metalworking techniques diffused from the Middle East to the Danube while bypassing the Aegean, the possibility that some of the techniques of metalworking were invented independently in both Spain and the Balkans must now be considered.

As Renfrew puts it, a 'fault line' has developed in the chronology of European history. The Near East, Turkey, the Aegean, Greece and Egypt all remain on the undisturbed side of the fault; the dates traditionally assigned to their stages of development, dates derived by historical methods, remain unchanged and are simply confirmed by radiocarbon dating, as recalibrated. But west of this fault line, though the relative chronology remains unchanged, the absolute chronology has moved, has slipped backwards in time as the result of recalibration of radiocarbon dates, so that now periods or 'phases' of European culture are contemporaneous with, quite different periods of Middle Eastern history. The temples of Malta now come before the temples of the Greek Bronze Age. The megalithic tombs of Brittany now go back to 4000 B.C., some 1,500 years before the pyramids of Egypt were built.

It does not follow from this that diffusion flowed the opposite way from Childe's way, from west to east. It would be ludicrous to pretend that the eastern civilizations learned from the western barbarians. The lesson is that the westerners were hardly barbarians at all and the origins of many of their cultural achievements are to be found in their own cultures. Renfrew writes:

The central moral is inescapable. In the past we have completely undervalued the originality and creativity of the inhabitants of prehistoric Europe. It was a mistake, we can now see, always to seek in the Near East an explanation for the changes taking place in Europe. Diffusion has been overplayed. Of course contact between prehistoric cultures often allowed ideas and innovations to pass between them. Furthermore evidence might easily emerge for occasional contacts between western or southern Europe and the Near East in early times. This however is not an adequate model for the explanation of culture change.

And he concludes thus:

The initial impact of the carbon 14 revolution will be to lead archaeologists to revise their dates for prehistoric Europe. This is the basic factual contribution that the tree-ring calibration has to make. . . The more profound impact, however, will be on the kind of explanation that prehistorians will accept in elucidating cultural change. A greater reluctance to swallow 'influences' or 'contacts' as sufficient explanations in themselves, without a much more detailed analysis of the actual mechanisms involved, is to be expected. This is in keeping with much current archaeological thinking. Today social and economic processes are increasingly seen as more important subjects for study than the similarities among artefacts. When the textbooks are re-written, as they will have to be, it is not the only European dates that will have to be altered. A shift in the basic nature of archaeological reasoning is necessary. Indeed it is already taking place in Europe and in other parts of the world. This is the key change that tree-ring calibration, however uncertain some of its details remain, has helped to bring about.[5]

Now this is what Binford says, exactly, when he announces the new archaeology in *New Perspectives*. Thus the new archaeology has come to Europe very largely as the result of the collapse of the entire diffusionist picture of European prehistory under the impact of radiocarbon and tree-ring dating.

The new archaeology does not reject diffusion as such; this remains a perfectly reasonably concept as explanation or part of an explanation of past cultural changes. What is rejected is the hypothesis that all European civilization can be explained by diffusion of ideas, techniques and peoples from the Middle East, and this hypothesis is rejected because testing against independent data in the shape of calibrated radiocarbon dates fails to confirm it. This process has been carried further in the very latest work on the origins of European agriculture.

At the Institute of Archaeology in May 1968 there was held a major international symposium on 'The Domestication and Exploitation of Plants and Animals', which was summed up by Stuart Piggott, Professor of Archaeology at Edinburgh University. It is a well-known question in examination papers on Neolithic Europe to ask students whether the truth lies in '*ex oriente lux*' or in '*le mirage orientale*'. Piggott says of this question:

An antithesis on paper, has it any solution in practice? It is in miniature (and perhaps on a larger scale than that) a crucial test of the diffusionist model of the relationship between the earliest agriculture in Asia and Europe, and our seminar has suggested that the light from the East is not an illusion, but may even be a fairly coherent beam. . . One thing does seem to stand up to the assaults of reasoned scholarships. With perhaps minor exceptions, Europe appears still to maintain its indebtedness for the initiation of agriculture to peoples, crops and herds from the lands across the Hellespont and beyond the Aegean. It is an old debt, standing now for some seven or eight thousand years.

That was said as recently as 1968, but by 1971, in a broadcast talk, Colin Renfrew was qualifying this view. 'Even in the case of the spread of farming to Europe – and the essential cereals (wheat and barley) were not locally available in Europe and had undoubtedly to be brought from the Near East – we are appreciating more and more that in each area the choice of crops and the farming system employed were a unique local adaptation to a particular environment.'[6] This is exactly what MacNeish suggested in 1964 – the multilinear development of agriculture, which he considered he could trace in the Valley of Tehuacan and its surrounding areas.

Excavations at Lepenski Vir in Yugoslavia have revealed a fully settled and permanent village dated to 5000 B.C. The theory of the Neolithic Revolution has associated the development of settled village communities with the beginning of agriculture, yet the people of Lepenski Vir had no agriculture and no domesticated animals except the dog. The point is that the village is on the banks of the Danube and the people lived primarily by fishing and hunting. There is evidence from Africa that village development started in some cases because of reliance on fishing and the necessity of having a riverside settlement, with agriculture developing through a 'cottage-garden' horticulture stage. Yet without looking beyond Europe, Renfrew can write:

> Other animals, notably cattle and pigs, may well have been domesticated in Europe in different places and at different times quite independently of what was happening in the Near East. We can now see that while the major cereal crops were brought to Europe from around the Fertile Crescent or from Anatolia, others, such as the vine, oats or flax, were locally domesticated. Many of the essential developments in farming were local ones.[7]

Perhaps that was just an hypothesis in 1971. Where is the independent evidence to test it, as the new archaeology demands? Some very surprising evidence appeared in January 1973, provided by Rainer Berger, Libby's former assistant, now Associate Professor of Anthropology, Geophysics and Geography at the University of California. Berger reported on 'Earliest Radiocarbon Dates for Domesticated Animals'. He used the method of dating bones directly by dating the collagen in them, rather than dating charcoal found at the same level as the bones; the collagen method was first developed by Libby and Berger in 1964 and has been improved since then. This is the first study of the earliest bones of domesticated animals from Near Eastern and European sites to use this method, and even further interest is added to it by the fact that Berger was able to work also on a good deal of material recently discovered by Russian archaeologists in the Ukraine and other areas of southern Russia.

Berger's conclusions are:

Our dates show that cattle and pigs were first domesticated in Europe. Sheep, which were thought to have become extinct in Europe during the terminal Pleistocene, also appear first in Europe. However there remains little doubt that sheep were first domesticated in the Near East or Turkey, since no wild sheep appear to have existed in Europe at the beginning of the Holocene [Author: modern times, geologically]. Dogs were domesticated in both the Near East and Europe at virtually the same time. In the Near East, Asiab, at around 8,000 B.C. qualifies as the first centre of goat domestication. It is also the earliest centre of domestication for all the animals we have dated here. Horses were first domesticated on the steppes of the Ukraine, perhaps even earlier than our dates indicate since all of the samples found at Polling were virtually contemporaneous. Undoubtedly future research will alter the details of our overall impressions especially after bones at earlier sites such as Nea Nikomedeia have been dated directly. But, on balance, there can be no doubt that South Eastern Europe was as much an early centre of domestication as the Near East was.[8]

It is necessary to add that Nea Nikomedeia is in Macedonia and the site that produced the earliest domesticated pigs and cattle was near by at Argissa-Magula – the dates for these bones are around 7000 B.C., slightly earlier than dates for the same animals at the now familiar Jarmo in Iraq and Cetal Huyuk in Turkey. Asiab, which produced the oldest domesticated bones of all, is a site in western central Iran – the familiar Zagros Mountains area.

It appears we must accept the idea of agriculture and cattle-domestication arising in many different places independently all over the world, dependent only on the local environment and the availability of potentially domesticable plants and animals. And in the broader field of archaeology and prehistory we must accept that we are in the middle of a revolution. The old paradigm has been cast out, the new archaeology may provide the 'new paradigm'. But no one, in the middle of a revolution, can safely foretell the end.

1. C. Renfrew, *Before Civilization*, p. 15.
2. L. and S. Binford *New Perspectives in Archeology*, Ch 1.
3. D. Brothwell and E. Higgs, 'Scientific Studies in Archaeology', *Science in Archaeology*.
4. L. and S. Binford, op. cit.
5. C. Renfrew, 'Carbon 14 and the Prehistory of Europe'.
6. C. Renfrew, 'The Revolution in Prehistory'.
7. ibid.
8. R. Berger and R. Protsch, 'Earliest Radiocarbon Dates for Domesticated Animals'.

16

The New Archaeology

What does a new archaeologist do that is different from what his predecessors have done? The clearest example is probably the excavation of the 'Broken K Pueblo' reported by James N. Hill, Assistant Professor of Anthropology at U.C.L.A.

'Broken K Pueblo' is one of the ruins typical of the semi-desert area of Arizona, a sprawling mass of masonry remains, the ruins of simple single-storey buildings. It is situated in the Hay Hollow Valley eleven miles east of the oddly named town of Snowflake, Arizona. The ruined buildings show that there were ninety-five rooms, all stuck together in two main blocks, but all irregularly shaped. In addition there are the remains of a few outhouses or outlying buildings. It was occupied at the beginning of the thirteenth century A.D., but present-day Pueblo villages occupied by the Hopi and Zuni Indians are constructed in exactly the same way and shape as this ruin of seven hundred years ago. The two main blocks of buildings roughly surround a central area which has remains of activities in it.

A site such as this is much too big to excavate completely; so the excavators contented themselves with drawing up the general plan of all the buildings and then concentrating their digging on some forty-six rooms. These were meticulously cleared out, with all the material being put through a mesh screen to ensure that nothing larger than a quarter-inch in size was lost. The central area was scraped down to the activity level with a mechanical scraper and great attention was paid to ashpits and rubbish dumps.

It was soon apparent that not all the ninety-five rooms in the sprawling complex were identical. First there were found four 'kivas', the ceremonial rooms of the Indian tribes of the south-western U.S.A. which are still found in modern Pueblo Indian life. Then it became clear that, broadly speaking, there were two types of room in the ruined pueblo – large rooms and small rooms. The large rooms usually had firepits in them and many of them also had 'mealing-bins' (for storage) and a ventilation apparatus. However, some of these items were also found in a few small rooms and not all large rooms had them. The archaeologists therefore measured all the rooms they had excavated with great care, and performed a statistical analysis on the size of the rooms. It was immediately clear that there was not a continuous spectrum of room sizes, ranging steadily from very small to extra large – there was instead a clear mathematical distribution into one group of twenty-six large rooms, a second group of twenty-four small rooms and the four special rooms or kivas. (By this stage a further eight rooms in addition to the original forty-six had been excavated.)

Obviously the original inhabitants had built large rooms for one purpose or set of purposes and small rooms for another objective, quite deliberately. But was it a case of large rooms for large families and small rooms for small families or did one family unit tend to own both a large room and a small room, and use the different rooms for different purposes? The most obvious answer would be that each family used the large room with its firepit and storage bins for food preparation, cooking and most of the activities of daily life, and the small, dark rooms as storage places.

So, in the usual way, Hill looked at the modern Pueblo Indians and also looked up the ethnographic work done among these tribes. Sure enough both Hopi and Zuni tribes in modern Pueblo villages have large rooms containing firepits, mealing-bins and ventilation systems which, although not identical to the prehistoric types, serve the same purpose. They also have small, dark, storage rooms attached to the large rooms. But, just as in the prehistoric ruins, the rules are not absolute; some modern storage rooms have mealing-bins or firepits in them although most do not.

The modern Indians also have another sort of room in their

villages, still larger rooms which they use as clanhouses. So the question arose as to whether the largest of the main rooms in the 'Broken K' ruin were clanhouses – and seven rooms were possible candidates for this function.

From all this it would seem reasonable to assume that the room functions in the thirteenth-century pueblo were the same as the functions of the similar rooms in modern Indian villages. But this is where the new archaeologist has something to add. He says that he must test this hypothesis and the deductions that can be made from it.

James Hill, in fact, set up sixteen propositions deduced from the hypothesis and testable against the archaeological record using only material which was independent of the facts used to arrive at the hypothesis, i.e. independent of the actual measurement of the rooms or the presence of firepits and so on. These propositions start like this:

1. If the large rooms were all-purpose living rooms they ought to contain larger numbers and higher densities of most categories of cultural remains than either the small rooms or the special rooms. If the small rooms were storage rooms they should have the lowest densities of most materials, with the exception of food crops, and the special rooms should fall somewhere in between.

2. The large rooms should contain a wider variety of materials than are found in the other room types, since the largest number of different kinds of activities were presumably performed in them.

3. The large rooms should contain evidence of the processing or preparing of food prior to cooking, independent of the mealing-bins...

Proposition 6 says that the large rooms should have evidence of water storage and therefore should have more remains than other types of room of large, narrow-necked, decorated jars. Proposition 7 says that large rooms should contain evidence of manufacture or craft work such as appropriate tools and debris. Proposition 9, however, states that small rooms should contain more remains of the large undecorated jars which were presumably used for general food storage in the way modern Indians used such jars. And proposition 10 states that the storage rooms should show evidence of the remains of other things as well as food, perhaps ceremonial paraphernalia and tools and such things as seeds for planting.

Propositions 12 to 14 set out the tests for the use of the special rooms like modern kivas. Proposition 15 says that if the largest of the large rooms were used as clanhouses they should have some ceremonial remains in them. Proposition 16 concludes, 'It can be expected that the three major kinds of rooms at the site did not all have the same context with regard to the sexual division of labour. The special rooms should contain cultural items associated with male activities primarily, while the large and small rooms should contain both male- and female-associated items.'

Virtually all these propositions are expressed in basically mathematical terms, that is they set up measuring standards, and can be proved or disproved by counting what has been found. So the next step, the actual testing of the deductions, consisted of a careful counting of what was actually dug up and correlation with the type of room in which it was found. The stone tools were divided into classes such as 'projectile points', various forms of knives, scrapers, choppers, axes and mauls, with the food-preparing and grinding stones called metates and manos forming yet other classes. Then the mean numbers of each type of tool found in each type of room were calculated. It was clear that the large rooms contained a greater proportion of all types of tool except antler-flakers. The small rooms contained particularly large quantities only of used flakes and blades and graver-burins. The special rooms contained the greatest number of projectile points, various 'male' working tools, like hammerstones, and also the ornamental items. Overall the message was quite clear: the large rooms had both the greatest number of artefacts found in them and also the greatest variety.

Similar analysis was then carried out with pottery types, animal remains, vegetable remains, including pollen, and a special class of pottery clearly associated with burial-procedures. The animal and vegetable remains were examined for evidence of cooking as well as simply for evidence of the place of storage

The sum total of this effort was that twelve of the sixteen propositions were fully confirmed by the evidence of excavation, and a further two propositions were partially confirmed. Two of the propositions (9 and 15) were not confirmed these two said respectively that small rooms should contain most of the remains of the undecorated 'storage' pots and that the largest of the large rooms should contain the remains of ceremonial paraphernalia if they were indeed clanhouses similar to the modern ones. Hill says of these results:

It is thus demonstrated that the three major room types at Broken K Pueblo were the functional equivalents of the three major kinds of rooms in present-day Pueblo villages; they may legitimately be called habitation, storage and ceremonial rooms (or kivas) . . . If the rooms were not functional equivalents of the modern room types, it is highly unlikely that so many of the expectations in terms of contents would have been confirmed. It has not been possible, however, to demonstrate that the seven extra-large rooms at the site were functionally equivalent to the modern clanhouses. It could not be shown by this analysis that they were different from the other habitation rooms, even in the terms of floor area; they appear to have been nothing more than large living rooms. Nonetheless they may have been clanhouses, and a more detailed study of the present-day clanhouses may suggest other evidence that can be looked for in prehistoric pueblos to document the existence of such rooms.

The failure to confirm the predominance of undecorated storage jars in the storage rooms, though a minor point, has some significance. It shows that the enthnographic evidence from modern Indians on this point cannot be projected into the past – that at some time between A.D. 1300 and the present there has been a change in the storage techniques. Therefore undecorated jars found in prehistoric sites cannot be declared to be necessarily storage jars. Hill concludes,

It is evident, then, that archaeology can contribute information of its own; propositions can be generated and tested, and a great many things can be learned that have not previously been known or speculated about before. In addition to making use of available ethnographic evidence, it is possible to discover new information that will be of use to ethnographers and social anthropologists. This general methodology of generating and testing propositions is, of course, not restricted to use in the American South West; it is universally applicable.

The substantive conclusions presented in this paper are not, of course, wholly new; investigators have frequently been able to identify different kinds of rooms in prehistoric pueblo sites. It is clear, however, that the functional equivalence of prehistoric and historic types has *not* previously been demonstrated. Functional equivalence has been *proposed* (and believed), on the basis of the fact that there are obvious formal similarities; but the proposition has not been adequately tested.

Still, I do not pretend to have presented many substantive conclusions that have not already been believed for a number of years. I have used this particular South Western example as a vehicle to illustrate what I think is an important methodological approach in archaeology. The substantive material was chosen because it is familiar to most archaeologists and is not in itself controversial.[1]

The new archaeology can, however, generate much more fresh information than Hill's purposely chosen 'textbook example'.

In considering the light which chemical analysis can throw on the sources of prehistoric materials (see Chapter 10), the work of Colin Renfrew and his colleagues in tracing the sources and trade routes of obsidian in the Middle East in Neolithic times was quoted. Now that work has been taken further by the question-asking, hypothesis-testing and statistical analysis methods of the new archaeology.

Chemical analysis of the obsidian showed that, as far back as excavation can yet take us, obsidian from one or two sources in central and eastern Turkey can be found in all the early sites of the Middle East, even though the quantities are very small. As early as 8000 B.C. Jericho was obtaining obsidian from these sources as much as five hundred miles away. This sort of discovery changed the general view of the efficiency and long-distance carrying-power of Neolithic trade. Likewise tracing the obsidian from the Greek island of Melos has shown that man must have been able to cross the sea and trade by boat or raft at least as early as 7000 B.C.

But this pushing back of the time-scale of trade has not been its main importance. Colin Renfrew says:

Prehistoric trade . . . now appears in a completely new light. Formerly its chief interest for the archaeologist was as a proof of contact, as a sure sign that goods, and with them no doubt ideas, were moving from place to place; an indicator, indeed, of these migrations and influences. Along these trade routes, it was thought, the great advances of civilisation were diffused. . . We are beginning now to see trade, not so much as a proof of contact between distant lands – which of course it often is – but as an indication of new social and economic relationships within the society itself. Several developments in archaeology, both owing much to the natural sciences, have revitalised our approach.'[2]

Renfrew carried his studies of obsidian trade routes a stage further by comparing the amount of obsidian found at different sites against the distance of the site from the source. This showed the surprising fact that the amount of obsidian at a Neolithic village fell exponentially according to its distance away from the source; that is, in non-mathematical terms, the villages nearest the sources had much the most obsidian, but as the excavations moved away from the source the amount of obsidian found at the sites did not decrease steadily but decreased very rapidly indeed until the furthest sites had only minute quantities of the precious material. This finding argues very strongly against the 'trade' in obsidian having been carried out by 'traders' carrying sackfuls of obsidian away from the sources and selling it in different villages. Renfrew puts up the hypothesis (he calls it 'controlled speculation') that the mode of transfer of obsidian – the method of trading – was that the obsidian was handed along from one village to another, each village keeping a proportion of what it received by 'trade' and using the rest to 'trade' with the next village down the line away from the source. In mathematical terms this mode of handling obsidian would satisfactorily account for the distribution of obsidian and the exponential decrease in the amount as the distance from the source increases. But this method of trade from village to village casts grave doubts on our ideas that trade was 'commercial', that

Neolithic trade was anything like our modern trade or even barter-systems.

It is well known among anthropologists who have studied living examples of primitive societies – South Sea Islanders or the natives of New Guinea for example – that it is very common to find a system of exchanging products between villages and communities and families which is not based on hard commercial bargaining, but is based on 'present-giving' instead. The object of this system is not profit-making, but the setting-up or cementing of a social relationship. Generosity, friendship, prestige, status, mutual defence and mutual support are much more important driving factors than making a good bargain in this system. Marshall Sahlins, the American anthropologist, has termed this system of gift-giving 'balanced reciprocity', and it is not unkown within our own modern hard-headed, money-based societies. The obvious analogies lie in Christmas presents, birthday gifts and wedding presents, where we usually wish to be seen to be as generous and kindly as our friends, neighbours and relatives, yet the richer among us have to avoid patronizing the poorer while at the same time they do not wish to appear 'stingy' with their wealth.

Such systems of gift-giving have often become highly formalized among societies which have lived in primitive conditions for hundreds of years, and in some places the gift-giving is the main basis for social relations. If such a system pertained among the villages of the Neolithic Middle East, it would have had enormous advantages in regulating contacts between communities and breaking down isolation as well as in distributing a valuable material such as obsidian. From the new archaeologist's point of view, the hypothesis that obsidian 'trading' was carried out in this way conforms mathematically with the discovered distribution of the material. The social hypothesis that such relations existed between the Neolithic villages of the Middle East remains to be confirmed by independent tests, though there is modern ethnographic evidence to support the idea that such a 'trading' system could have existed in the past. This, say the new archaeologists, is the way in which other sciences, the natural sciences, have progressed so far and so fast – a good hypothesis is one that can be tested and which sets up valuable and interesting questions in the process.

This approach to archaeology, in which a counting and measuring of objects found is used to validate or invalidate an hypothesis, is held by the 'new archaeologists' to be the 'new paradigm', the new framework of thought within which work is done. The old arguments about whether diffusion or evolution have caused the changes that can be seen in the archaeological record is seen not so much as wrong or right but as irrelevant.

It was the demolition of the old view of European prehistory under the impact of radiocarbon dating that caused the chief doubts about the 'old paradigm'. Colin Renfrew writes:

At first sight the sudden collapse of the diffusionist framework for Europe might tempt us to revert to an evolutionist viewpoint. But this is no longer adequate. For to ascribe all progress everywhere to innate properties of the mind of Man is to give an explanation so general as to be meaningless. Unless we have some independent means of discovering these properties, and of testing whether they do indeed govern cultural changes and developments, the theory says little more than that everything that has happened in human history is to be seen as the outcome of man's mental structure. This view is about as helpful as the proposition advanced by Voltaire's Dr Pangloss, that we live in 'the best of all possible worlds'.

In order to disentangle ourselves from this old and arid debate, it is sufficient to see that 'evolution', applied to human culture, need imply little more than gradual development without sudden discontinuity. We would all agree, moreover, that ideas and innovations can be transmitted from man to man and from group to group, and that this is a fundamental distinction between biological and cultural evolution. All this is perfectly acceptable, but it does not supply us with any useful or valid explanatory principle.

In rejecting both evolution and diffusion as meaningful explanatory principles, we are rejecting much of the language in which conventional prehistory has been written. For both localised evolution and more general diffusion were essential components of the first paradigm, the general language and framework of the prehistory built up in the century following the publication of Darwin's *Origin of Species* in 1859, and the demonstration of the antiquity of Man – that man has a prehistory extending back far before the Biblical Creation – in the same year.

The old order, then, is changing, and the task of the New Archaeology today is to construct a more effective way of speaking about the past, a new language implying fresh models of the past – a new paradigm. Until we can do this, we shall simply have shown up some shortcomings in the traditional view without demonstrating that something better might be put in its place. And, as the Americananthropologist Julian Steward remarked, 'Fact-collecting of itself is insufficient scientific procedure; facts exist only as they are related to theories and theories are not destroyed by facts – they are replaced by new theories which better explain the facts. Therefore criticisms . . . which concern facts alone and which fail to offer better formulations are of no interest.'[3]

But it is by no means clear what form the new paradigm will take. Some hold that models should be taken from the strict, physical sciences. Others say that there is no such thing as archaeology-science, there are only techniques working in aid of anthropology. The impact of the laboratory sciences in destroying the old paradigm is likely to carry through in the formation of the new. But the representatives of the 'humane studies' point of view continue to voice their views, and few archaeologists are likely to be happy about any threat to divorce them totally from the traditional activities of scholarly historians. Indeed, in view of the latest developments in archaeology, which are linking the archaeologist with the historian, by using archaeological techniques to supplement the evidence of documents (see Chapter 17), it is likely that archaeology will continue to be closely related to 'historical' schools, though the relationship will undoubtedly be different from the one that pertained ten years ago.

It was Renfrew who organized the first big International Conference of the new archaeology in Europe; it was called a Research Seminar on Archaeology and Related Subjects, and the official subject of this three-day meeting at Sheffield University in December 1971 was 'The Explanation of Culture Change: Models in Prehistory'. The make-up of the conference and the subjects discussed reveal much about the new archaeology.

A considerable proportion of the time was taken up with the discussion of the nature of scientific processes, the use and abuse of models, using that word in the sense of a scientific hypothesis-generating concept, and the general nature of laws, theories and epistemology. Professional philosophers were as prominent as archaeologists in this part of the seminar. Ethnographers, anthropologists, zoologists and botanists, linguists and ecologists, were also there to present papers. Computer experts and mathematicians mingled with the archaeologists and the many American archaeologists who are officially anthropologists. Opponents of the new archaeology such as Jacquetta Hawkes were there, too, though in a minority. The American contingent included many of the people mentioned in this book such as Lewis Binford and Frank Hole. A large contingent from France, including the famous François Bordés, the expert on French Stone Age tools, showed that the ideas of the new archaeology were not confined to English-speakers, and there was even a paper from the University of Leningrad. Denmark, Italy, South Africa, Czechoslovakia and Australia were also represented.

While some sessions were redolent of the vocabulary of the new archaeology (jargon, the opponents label it) such as 'Multivariate Analysis of Change in the Culture System' or 'The Explanation of Artefact Variability in the Palaeolithic', other sections had more urbane titles such as 'Movement, Trade, and their Consequences'. Individual papers showed the same sort of contrast in their titles, varying from 'On the reflection of cultural changes in artefact materials, with special regard to the study of innovation contrasted against type stability' to 'The safety pin and fashion in prehistoric Europe'.

The depth to which mathematical or quantitative thinking, which has previously been a characteristic of the 'harder' sciences such as physics, has penetrated into the new archaeology is well illustrated by one of the papers at this conference. In the section on the strictly archaeological subject of 'Problems in Data Collection and Data Processing', David H. French, Director of the British Institute of Archaeology at Ankara, pointed out that there could not be valid quantitative comparisons between material dug from two different sites unless it could be established that the same proportion of the same materials with respect to the total amount of material had been recovered. In other words, he seriously questioned the comparisons between amount or weight of obsidian recovered from various Middle Eastern sites and their distance from the sources of obsidian simply because Renfrew could not prove that the standard of data-recovery from one site was equal to the standard of recovery at any other site. French wants to see definable, measurable and repeatable methods of data-recovery brought into archaeological digs, just as a physicist might insist on knowing the precise machinery and measurements used in someone else's experiments. Unless a set of results can be obtained repeatedly by an independent experimenter, the physicist is not happy to include the original data in the general corpus of knowledge.

David French is directing a large current excavation programme at Asvan (pronounced Ashvan) on the upper reaches of the Euphrates, and in many ways this is a typical new archaeology project. In one sense it is a rescue operation, for the area is due to be flooded by the waters of a new reservoir the Turkish government is building. But it is also a project in 'regional' archaeology where several sites in a region are being dug, including an inspection of what lies below the present-day village of Asvan. None of the sites is expected to produce dramatic temples or palaces of lost civilizations, but the region appears to have been inhabited from at least the fourth or fifth millennium B.C.; there were Roman inhabitants when this was the local frontier of the Roman Empire, and there are clearly medieval remains as well as the modern systems of agriculture and life. The official report on the first stages of the project says that it was

> . . . developed largely in order to meet the changing theories and methodology of modern archaeology, but partly in order to set up a deliberate experiment in project work with a modern theme, i.e. work carried out by interested disciplines to a stated theme which would act as a cohesive element. The project was not intended as interdisciplinary but as interrelated work. The theme is thus a framework

within which the collection of data can be pertinently and economically co-ordinated. This approach is a logical extension of the view that information is largely collected within pre-existing, pre-conceived structures or frameworks.[4]

The theme of the project is the occupation, exploitation and manipulation of the environment of this region by man. The team at the site, as well as archaeologists, includes a geographer, an ornithologist, a botanist, an agriculturalist and a group of four sociologists and anthropologists to study the modern villages and the household activities of the inhabitants.

The whole project is, then, a variation, in tune with the latest thinking, on the sort of ideas with which MacNeish started the exploration of the Tehuacan Valley. But, where he was seeking by archaeological methods to discover the facts about the origin of cultivated maize, and was led to the spot by environmental considerations, at Asvan the object is to trace, by archaeological methods, the development of the environment under the impact of man and his activities, including agriculture. In addition French has set up an 'excavation strategy' which specifically aims at establishing 'mechanical, repeatable processes in the collection of materials and objects'.

Excavations such as the Asvan project demonstrate the continuity of archaeological practice under the influence of new thinking such as 'environmental archaeology' and the new archaeology – perhaps some archaeologist of the future will examine the differences between Tehuacan and Asvan as an example of 'culture change within archaeology'. But an entirely new field of exploration has recently been opened up by Binford and his colleagues.

Apart from his unofficial leadership of the new archaeology movement, Lewis Binford owes his high professional reputation mostly to his reinterpretation of the French Stone Age culture which is called Mousterian, after the rock-shelter at Le Moustier in which it was first identified. The essence of this reinterpretation was the application of the 'tool-kit' concept to the assemblages of Stone Age tools and the explanation of changes in these assemblages as being the result of different or changing activities for which the stone tools were used. He viewed the archaeological record as a by-product of behaviour. The changes which he could measure in this section of the archaeological record might not only be caused by changes in culture (caused by new peoples or new ideas diffusing into the region) but might just as easily be caused by changes in the environment stimuli to which people of the culture reacted by changes in behaviour.

What stimuli do cause changes in human behaviour of such a type as would appear in the archaeological record? Binford has recently been out to study a group of Eskimos who are still dependent for their livelihood on hunting the herds of caribou that roam the mountains of the Brooks Range in Alaska. For four years he has spent a considerable part of his life with one of the few remaining groups of humans who are still dependent almost solely on hunting for their existence. This study has been particularly aimed at finding out how the behaviour of these Eskimos changes as the environment changes with the seasons, and with finding out how the archaeological record of the future is being built up at the present time. What the anthropologists and archaeologists found came as something of a shock. Instead of a simple relationship between the Eskimos' tools and their activities at the different camps they use at different seasons of the year, there was something much more sophisticated. In Binford's words,

> I went to this group of people with the notion that variability in the archaeological record could be referrable directly to variation in the activities which these people performed at different locations in their environment. And what I found, much to my surprise, is that that was not correct. Why? Because this particular group of people have a highly efficient technology in which they curate their tools.[5]

What this means is that if an Eskimo has a favourite tool he takes great care of it; he mends it when it is broken; he carries it about with him wherever he goes and he takes it back home with him after he has used it. If he breaks it while he is working with it, oes not throw it away, he takes it back and tries to mend it at leisure, or to use it for something else. He recycles his raw material, in current terminology. If a tool is finally dropped somewhere so that an archaeologist of the future might find it, it is far more likely that it has been lost on a journey in just the way that a modern westerner will lose his umbrella or his cigarette-lighter. Binford summed up his experiences:

So what turns out then if one looks over a long period of time at the items discarded by these people, the frequencies with which the various items are discarded are in no way related to the importance of those items directly. In fact they are related inversely. Those items which are unimportant, which are expendable, are the most commonly discarded; those items which are most important in their daily lives are the items which they tend to curate, they tend to recycle. So that from an archaeological perspective the particular tools that were most important to their way of life tend to be the least frequent in the archaeological records. The tools, on the other hand, which were the least important turn out to be the most common items in their archaeological sites. This is essentially just the opposite of what archaeologists have always assumed. They've always assumed that the most common item is in turn the most diagnostic item and presumably then the most important item to the people in the past.[6]

Binford is careful to point out, however, that this conclusion based on present material cannot necessarily be applied directly to past behaviour. Apart from the theoretical qualification, there is, perhaps more important, the point that Eskimo technology is highly efficient in relation to their environment and objectives. There is every reason to believe that prehistoric technology was less efficient and that therefore it is more likely that prehistoric man made and used and discarded the tools he needed for a particular job in the place where he carried out that job. But the warning of Binford's latest work is there, nevertheless.

The new archaeology, then, is not simply a matter of applying computers and chemical analysis and plant-genetics to the data revealed by digging. It is an attempt to apply the quantitative methods of the natural sciences to the science of archaeology, not only in obtaining answers, but also in asking questions; not only in assembling data for the archaeological record but also in deciding what data to assemble. All the results so far obtained by applying the new scientific methods or by applying the new scientific thinking tend to show that primitive men, prehistoric men, the barbarians or the savages, call them what you will, were a good deal more creative, more inventive, more skilled and more efficient than ever we had thought.

1. J. Hill, 'Broken K Pueblo'.
2. From a broadcast in the B.B.C. series 'Rewriting Man's Prehistory', 2 December 1971.
3. C. Renfrew, *Before Civilization*, pp. 18–19.
4. D. French in the *Report of the English Archaeological Institute in Ankara.*
5. From a broadcast in the B.B.C. series 'Rewriting Man's Prehistory', 16 December 1971.
6. ibid.

17

Archaeology and History

History begins with the written record. In different parts of the world and for different peoples history, therefore, begins at different times. Archaeology has always been pre-eminently concerned with prehistory; the archaeologist produces the facts and interprets them for the reconstruction of the lives and stories of those who have left no written records.

Obviously the boundary line between history and prehistory is blurred and indistinct. The reconstruction of what happened in Roman Britain is partly based on the writings of Roman authors, partly based on the discoveries by archaeologists of sculptured inscriptions on tombstones, altars and the like, and partly based on purely archaeological work. Similarly, the reconstruction of the Maya civilization of Yucatan and Guatemala is partly based on pure archaeology and partly on the translation of the Mayan glyphic writing sculptured on stelae and temples.

But very recently it has come to be realized that archaeological methods of recovering data from the past can greatly increase our knowledge of what occurred in strictly historical periods – that the evidence of the spade can vastly increase our knowledge of societies and places which we have previously known only from written records. The most obvious place in which this can be done is among the cities of Europe which have been the sites of man's dwellings right back through history into prehistory. Doubtless it will also one day be possible to apply archaeology to illuminate the history of many places in, say, China or India. But it has also been shown that archaeology can illuminate such obscure historical areas as the impact of Europeans on the Indians of the north-eastern U.S.A.

This use of archaeology in helping the cause of history, as well as in its traditional role of revealing prehistory, is one of the most exciting and expanding features of the discipline.

It is impossible to say of such a development, 'it started exactly here at exactly such and such an hour' – this is sure to be unfair to some pioneer or even to many such. Let it simply be said that the attempt to study the history of Winchester in the English county of Hampshire and to provide archaeological evidence covering the whole of the city's two-thousand-year history began in 1961, because it had been decided to build the new Wessex Hotel on a car park just to the north of the cathedral. Local antiquarians believed that this site might be of great importance archaeologically. The man invited to direct the excavation before the hotel building could 'sterilize' the site was Martin Biddle, who has been described as 'the originator of urban archaeology'. The archaeological study of ancient Winchester continued for ten years after that, taking advantage of virtually every redevelopment that occurred anywhere near the city centre, and digging in a good many other places as well.

Winchester is right at the core of English history. It was the fifth largest of the Roman cities of Britain; it was the capital of the Saxon kings of Wessex who became the first kings of England; and although always of less economic importance than London, it was the place where many of the early Norman kings were crowned. It has the great Norman cathedral, one of the glories of eleventh-century Europe, but in Tudor times it started to go into a decline and became a very small country town by the nineteenth century. It has recovered, to become a bustling county centre though still relatively small in the twentieth century. For reasons quite unknown Winchester is connected with the legend of King Arthur; Sir Thomas Malory held it was Arthur's 'Camelot' and in the cathedral hangs an alleged 'Round Table' for Arthur and his knights, which is certainly not genuine but may be as old as the thirteenth century. At the other end of the scale, I.B.M.'s British research laboratories have been built just outside the city.

The excavations at Winchester have, perhaps because of the

historical romance of the city, captured public imagination in Britain and the rest of the world in a way that few modern digs have done. At least two American universities are officially associated with the digs (Duke and North Carolina), and many of the experienced archaeologists working on the sites have been American. Yet the digging has mostly been done by volunteers, among them a journalist, Jane Davidson, who subsequently wrote about her experiences in the *Guardian*; she began her description thus, 'I have just spent a week in the nearest thing to a labour camp that Britain has to offer', though she pointed out that all the inmates were volunteers!

Digging, she found, was not so easy: 'Within an hour both wrists ached, blisters were coming along nicely and my backside felt raw from sitting on stony ground. Also I was none too certain what I was doing. "God!" exclaimed Robert [the supervisor of her area, whom she described as 'an ebullient redhaired American anthropologist'] when he came to check my progress, "Stop hacking mindlessly at that medieval brown layer."' But later things settled down: 'Digging involves discovery not only about strata, but about oneself. Initially as I relentlessly scraped, I became somewhat gloomy by mid-morning. Then I reflected that never normally do I perform any continuous action; my time is punctuated by small dramas which preclude any sustained thought. Gradually the inevitability of digging became very calming.'

Her fellow volunteers included a Japanese student, a Southampton social worker, a Croydon businessman and a large group of Americans, some seeking experience, others doing fieldwork towards their degrees. Jane Davidson adds, 'Diggers are instantly distinguishable from the tidy Winchester townspeople by their suntanned faces and mudcaked blue jeans with a trowel sticking out of the hip pocket. They have a notable impact financially on small shops and the town's 72 pubs.' Throughout the ten years of the Winchester excavation digging was almost entirely confined to the ten weeks of the summer holidays.

The excavations[1] have eventually revealed that Winchester began as a late Iron Age small hillfort and enclosure, probably occupied by the remnants of a tribe which had not been swallowed up in the invasions of the Belgae. (The Belgae were Celtic tribesmen who crossed from Gaul in the last centuries B.C.) The Romans, however, called their town on the same site Venta Belgarum, and its importance was largely as a centre of communications. The Roman forum and the remains of some Roman shops and houses were found in the very first excavation in 1961 on the hotel site near the cathedral. In the next four years the Roman street plan was reconstructed based on the findings of carefully gravelled Roman road-surfaces. But on these Roman road-surfaces later houses had been built either by the last of the Romano-Britons or by the first of the Saxons. Whoever built them, 'they are evocative of a ruined town in which civic order had been abandoned and where the streets offered the best-drained, rubble-free foundation for a home'.

It had always been assumed that the modern street plan of Winchester – a grid pattern based on an east-west-running High Street – was based on the original Roman streets. The first surprise of the Winchester excavations was the finding that this was wrong. The street plan is Saxon. It is known from historical records that the Saxon Gewissae regarded Winchester as their capital in the seventh century, and at this time King Cenwalh of Wessex moved the episcopal see from Dorchester-on-Thames to Winchester and built a Saxon cathedral there. By digging in the traditional manner, guided by electrical resistivity surveys, the foundations of the Saxon minister were found, though the Normans had stolen almost all the stone for their eleventh-century cathedral. Later this Saxon building was enlarged and rebuilt several times, particularly as a fitting shrine for the bones of St Swithun, a Saxon bishop still remembered in Britain because of the old myth that the weather on St Swithun's feastday is a good prediction for the weather conditions over the following forty days.

There are one or two Saxon charters still preserved from the tenth century, relating to property transfers, that show that the present streets of Winchester were in existence by A.D. 990. In 1964 it proved possible to dig below the medieval surface of a street on the line of the present-day Trafalgar Street. Five successive cobbled surfaces were revealed and below the lowest lay a silver penny of Edward the Elder, the son of King Alfred. On the surface of this street was found a silver dirham, a coin minted in Samarkand in A.D. 898. The Saxon streets were not laid above the Roman streets, which had had houses built on top of them in the days after the departure of the Romans in 407. The Saxon streets were therefore laid out on a grid plan, to which modern Winchester conforms, as early as the first years of the tenth century; and the significance of this is that a number of other English county towns, such as Oxford and Exeter, were probably laid out at about the same time, when King Alfred and his immediate successors rebuilt after their victories over the Danes.

The dating of the Winchester street-plan to this period takes with it the dating of a number of other southern English (Wessex) town-plans, all showing clear signs of a regular grid-pattern. There was thus in late Saxon England a group of regularly planned towns, an observation that must cause considerable revision of current views about the extent and nature of Anglo Saxon towns and royal involvement in them.

But there were even more important results to come.

Perhaps the most important single result of the excavations and their associated research has been the recognition of the Old English Royal centre which gives Winchester its clearest claim to be regarded as the forerunner of Westminster and in some senses the Old English capital. The grouping together in the South-East corner of the city of the Old Minster, the cathedral church of Wessex, and burial and coronation church of many of its kings, the New Minster, the burial church of the Alfredian dynasty, and the Nunnaminster – all three royal foundations – together with the existence of a late Saxon bishop's palace and the probable, if not certain presence of the Anglo Saxon Royal Palace immediately west of the Old and New Minsters, forms a cluster of major structures of the first importance and without parallel in Anglo Saxon England. Only perhaps in the Carolingian world, whence late Saxon Winchester drew so much of its artistic inspiration, can a parallel be found in Northern Europe and that parallel is Aachen. [Aachen was Charlemagne's capital: Author].[2]

Excavations showed that the Normans took over the lot; they rebuilt the cathedral, they rebuilt the bishop's palace and the royal palace and they constructed a new castle, at first mostly of earthworks, on the western hill to dominate the city. Winchester, however, continued in much the same way as before only under new masters; the excavations show quite clearly that there was no gap in the buildings of the ordinary citizens when the Normans came. The earliest documents confirm this, for in such things as the names they gave to their children at their christening it was many years before the babies were called by the new Norman French names.

It has been in elucidating the lives, the life-ways, of the ordinary citizens of Winchester a couple of centuries later, that the Winchester excavations have shown the greatest significance for both archaeology and history. The chance came in 1962 when a new post office was to be built in the Brook Street area of the city. The clearance of the site allowed the diggers to get at a whole series of ordinary shops, workshops and dwelling houses stretching from the fifteenth back to the thirteenth centuries. Brook Street is a very muddy area – the water-table is very near the surface – and the water in the ground had preserved the wooden foundations of the medieval buildings in rather the same way that Danish bogs preserve so much organic material. They found a row of little cottages, apparently the remains of the houses of the lowest-paid workers in the textile industry. There were the houses of the rather-better-off fullers and the stone-built houses of much richer men, merchants who both dealt in cloth and carried on some of the manufacturing processes in their own houses.

To identify the work carried out in a medieval cottage as dyeing or fulling would be almost impossible from the archaeological record alone at the first time of asking. This is just where the combination of archaeology with traditional documented history comes in. The surviving documents of medieval Winchester include leases and wills, enrolment books, accounts and surveys, which have made it possible to trace who owned which property and when, from the fourteenth century at least, and in many cases much earlier, right down to the present day. The archaeological finds, in fact, could be related to actual people with known names, who had lived and worked in these houses. Furthermore, the occupations of these individual owners of archaeological data could be found, and hence the detailed layouts of the little workshops could be definitely connected with the various processes of dyeing and finishing the cloth. In the future this information can, therefore, be applied to other archaeological findings where there is no documentary evidence of the particular trade of a medieval occupant of a house.

These medieval textile workers lived in wooden houses with a long and narrow room on the street frontage which served as both shop and workshop in many cases. In the eleventh century they usually had a larger rectangular room with a central hearth and fire as a family living-room behind the shop. The houses were entirely built of wood and were single-storey. Only from the twelfth century onwards did stone start to be used, and then usually by one of the richer men building a hall over his garden at the back of the house. It was not until the thirteenth century that most of the buildings were of stone or at least had their foundations and wall-bases built of stone.

The bringing together of archaeological and documentary evidence presented several problems, however, and Biddle writes of them:

The documents are primarily evidence of the state of mind of the person who wrote them; they record for the most part political, legal, financial and administrative matters, both secular and ecclesiastical. They do not record the facts of daily life, the plan and character of the town, of its streets, houses, shops, palaces, churches and defences; they do not record, except incidentally, the historical processes of origin, growth and decline as these would appear in the changing fabric of the town. Such things were the setting of daily life; there seemed no reason to record what any man could see. Archaeology provides primary evidence for precisely this material setting and physical evolution, but it cannot normally add to our understanding of personalities and politics. It is only through a combination of these two complementary sources, documentary and archaeology, that a balanced account of urban history can be written, but the problems involved in such a combination are formidable. These problems arise from the fact that the evidence of archaeology and of documents is evidence of different kinds, the one material, the other abstract. Examples of the kind of question that would have to be faced in practice could be stated thus: is archaeological site A the same as parish or unit a in the documents? If so is the house A1, archaeologically defined, the same as property a1 in the records? If it is, does the archaeological phase 10 of house A1 represent the tenement on property a1 mentioned in 1340? . . .

Other problems arise from the fact that few British historians have archaeological training, and few archaeologists are documentary historians. The reason for this is the comparatively recent recognition of mediaeval archaeology as a discipline of importance in historical studies, and the result has been an unwillingness on the part of historians to use archaeological evidence, the value of which they felt themselves unable to estimate. Archaeologists on their part often ignore or uncritically accept documentary evidence.[3]

The excavations at Winchester have now stopped, largely to give a chance to publish the work so far completed. At this stage, work in the laboratory has taken over from digging and there is much hope of new information and new types of information. Thousands of animal bones have been recovered, for instance, and it is hoped that from them will come not only information on the diet of medieval Englishmen, but also information on the breeds and types of cattle being used in England in medieval times. There is also the work on the health of the inhabitants of Winchester (mentioned in Chapter 12). In simple, but not exaggerated, terms, the Winchester excavations have opened up enormous new horizons for future co-operation and mutual enrichment between archaeology and history.

In the north-eastern United States 'history', in the written sense, begins in the seventeenth century with the arrival of the British Pilgrim Fathers, the Dutch in New Amsterdam (New York) and the French in Canada. In the early history of this part of the American continent, in the seventeenth and eighteenth centuries, no name is better known than that of the Iroquois. The Iroquois helped the English and their colonists to defeat the French and their colonists with their Indian allies the Hurons, and thus to capture Canada. But the history of the Iroquois was short, for they fought on both sides in the War of Independence and were virtually destroyed by the American Revolutionary armies by 1780.

There were five tribes in the original Iroquois Confederacy, called 'The Great League of Peace': the Senecas, the Cayugas, the Onondagas, the Oneidas and Mohawks. At the height of their power, shortly after they had first met the Europeans and when European guns and other trade goods filled their camps, the Iroquois controlled an area from Lake Ontario to the Tennessee River and from Illinois to Maine. Francis Parkman, the greatest of nineteenth-century American historians, called them 'the Romans of the New World'. Naturally, therefore, the Iroquois play a great part in early American 'history', using that word in its academic sense as meaning documents and written records. Yet there were probably only 12,000 people in the entire Confederacy and they could raise at best little more than 2,200 fighting men. They knew of farming and they cultivated maize, beans and squash plants. But much of their subsistence came from the men's hunting and fishing and the women's gathering of wild plant-food. They had no domesticated animals except the dog. The question obviously arose as to why they were so influential and it has been stated that more ink has been spilt over the Iroquois than over any other tribe of American Indians.

Because the Cherokees adopted European ways more readily than most Indians, they were considered to be 'superior' and 'more civilized' than other Indian tribes. They also spoke a language akin to Iroquois, so some early scholars suggested that the Iroquois originally came from the Cherokee homeland in Georgia and the Carolinas. But opposing opinions brought them down the St Lawrence from the north and even from the Pacific north-west coast. Within the last ten years archaeology has settled the problem and shown that the Iroquois began just where the Europeans found them. In particular it has been shown quite clearly that the Onondagas, the central tribe of the confederacy, the tribe that was 'keeper of the central fire', had lived in an area of no more than twenty-five miles by fifteen miles in upper New York State ever since A.D. 1000. This home area of the Onondagas was near the modern city of Syracuse, so it is hardly surprising that most of the archaeological work has been done by teams from Syracuse University. The moving spirit in most of this work has been James A. Tuck, originally a botanist, who turned to anthropology while he was still at Syracuse in the late 1960s, and then became Assistant Professor of Anthropology at Memorial University in Newfoundland.

The discovery of prehistoric Indian sites in this part of the world is difficult because all that remains of their dwellings are the post-holes in the ground, the places in which they stuck the saplings that formed the palisades of their villages and the structure of their bark-covered houses. These post-holes are technically called post-moulds because all that remains is the dark-brown humus filling of the original hole in the subsoil under the modern cultivation. But, once found, the sites contain the usual debris of stone tools, pottery, vegetable remains, and even clay pipes for smoking tobacco. Later sites show clearly that European trade goods arrived before the Iroquois had met the European. Eventually the 'native culture' of the Iroquois almost completely disappeared. Shortly before they were destroyed the Iroquois used virtually nothing except goods of European origin.

In technical, anthropological terms the Iroquois rose from the Late Woodland Indians of the Owasco culture. In 1959 a team from the New York State Museum at Albany, under William A. Ritchie, excavated a site near Skaneatles Lake, called the Maxon-Derby site, in which there were the first foreshadowings of later Iroquois forms among the arrow-points and bone tools of an Owasco cultural complex. This village was dated to A.D. 1000. In 1963 another site was excavated by Ritchie, only a few miles away from the first, and called the Kelso site, showing a fully developed early Iroquois culture dated to 1390. Later work in the same area by Tuck has revealed other sites that fill in the gap both in terms of time and in terms of the development of a culture that can be called Iroquois out of the previous Owasco culture. In other words, Iroquois Indians had existed in roughly the area in which the Europeans found them ever since 1400, and archaeology shows that their culture had developed in that same homeland in the preceding four hundred years.

Nearly thirty more sites have since been excavated in the area immediately west of Syracuse, between Lakes Skaneatles, Onondaga and Cazenovia, which provide a steady series of dates from the end of the fourteenth century through to historically recorded Iroquois villages of the eighteenth century. They show the steady development of the culture that Europeans described when they first met the Iroquois. Whereas the first villages in A.D. 1000 were not defended, there were triple palisades of saplings round the first truly Iroquois village of 1390. By this time, too, there were the first longhouses. This typical feature of Iroquois culture was built in the normal Indian style of bark on a framework of saplings, but might well be hundreds of feet long, twenty-five feet wide, with a series of hearths down the middle (in later examples) which were apparently shared by related families living in 'compartments' stretching the length of the structure. There might be three or four longhouses inside the palisade of one village, and the steady growth of the population is shown by the regular increase in the size of the longhouses, until one more than four hundred feet in length is found in a village dated to 1410 by radiocarbon dating. Furthermore, the closeness of groups of villages each giving successive dates suggests that the Iroquois communities moved to a near-by site every twenty or thirty years, probably because they had exhausted the fertility of the earth and the local supply of timber.

But then the excavations turned up a surprising new development. Two villages, each showing slightly different traditions in their pottery-making and other artefacts, are found within two miles of each other and giving the same dates of occupation – from rather before 1450 to just after 1480. Tuck suggests that this grouping, which must have been based on some mutual non-aggression pact, although the villages were heavily defended, implies the start of what was to become the Onondaga tribe. At the time the Europeans found the Iroquois each tribe lived in a group of two or three villages situated near to each other. From this point onwards the archaeologists can trace that successive sites are in pairs of roughly contemporary villages.

The tradition changes slightly in other ways about this time. The villages were not only well fortified, they were usually placed on steep rises as well. The houses within the palisades tended to become more numerous and distinctly smaller than the vast old longhouses – and this may represent a social change as well.

There is a tradition, locally, that Europeans first contacted the Iroquois properly (that is apart from possible unrecorded traders' contacts) at a place called Indian Hill by a little stream between Limestone Creek and Butternut Creek high in the Alleghenies, east of Syracuse, when French missionaries started the work of trying to christianize them in 1654 and 1655. Certainly Father Simon LeMoyne records contacting the Onondagas on 5 August 1654. He says he was 'singing the ambassador's song', and that he received 'addresses of welcome' from the Indians – but it is not clear where this all happened or where the two priests, who came the following year to build a bark chapel, did their work. Jesuit records disagree with the local tradition, and they date the settlement of the Indian Hill site to 1663. The archaeologists can point to the sites they have found that seem immediately to have preceded Indian Hill – a linked pair of sites called Carley and Pompey Center. From Father LeMoyne's description it seems it could have been either of these that he entered.

An independent European documentary description of Indian Hill comes from 1677: 'The Onondagoes have but one town, but it is very large, consisting of about 140 houses, not fenced; it is situated upon a hill that is very large, the back on each side extending itself at least two miles, all cleared land whereon corn is planted. They have likewise a small village about two miles beyond that consisting of about 24 houses.' (The spelling has been modernized.) Tuck comments:

The visitor's words attest to the Onondaga two-village settlement pattern and also to the extensive plantings that surrounded the settlements. It further allows us to deduce that the dwellings at Indian Hill were not the traditional Iroquois longhouses; the site is much too small to have accommodated 140 houses of any great size. That the settlement at Indian Hill was 'not fenced' is surprising. Evidently by this time the Onondagas no longer built the customary village palisades.[4]

The archaeologists believe they have established the site of the smaller village, however, at a near-by spot called Indian Castle. Both archaeology and documentary evidence record the later movement of the Onondagas to sites called Valley Oaks and Toyadasso. (After the decimation of the Onondagas in 1779 the survivors first went to a reservation near Buffalo and they finally moved to the Onondaga Indian Reservation near Syracuse.)

The significance of this work lies, at least as far as this book is concerned, in the linking-up of archaeology and documentary history, so that at a crucial period one set of data illuminates, corrects and complements the other. The interpretation of both data-sets can provide us with a better reconstruction of the life-ways and culture-changes of the past. If this conjunction of disciplines can elucidate the growth of an historic British city and the clash between native Indian and invading European in north-east America, then it should have an ever-expanding role in historical and archaeological work in many other parts of the world for many different periods of time.

Binford wrote, when advocating the new archaeology, 'If we are successful . . . we will see an expansion of the scope of our question-asking which today would make us giddy to contemplate.' Archaeologists have not been turned giddy by the several different revolutions which have affected them in the course of the last twenty years, and which I have reported in this book. But if archaeology, and the new archaeology in particular with all its scientific alliances, are to expand on a large scale into those areas which have previously been the preserve of 'historical studies' then the prospect is giddying indeed.

1. The description of work at Winchester comes from a series of articles by Martin Biddle, the Director of the Excavations. I am also grateful for a personal interview with him.
2. Biddle, 'Winchester 1961–68'.
3. Biddle, *Archaeology and the History of British Towns.*
4. J. Tuck in *Scientific American.*

Appendix, Bibliographies, and Index

Appendix

Tree ring calibration curve
The relationship between calendar
dates and radio-carbon dating, basec
Suess of the University of Californi:

Conventional radiocarbon dates in radi

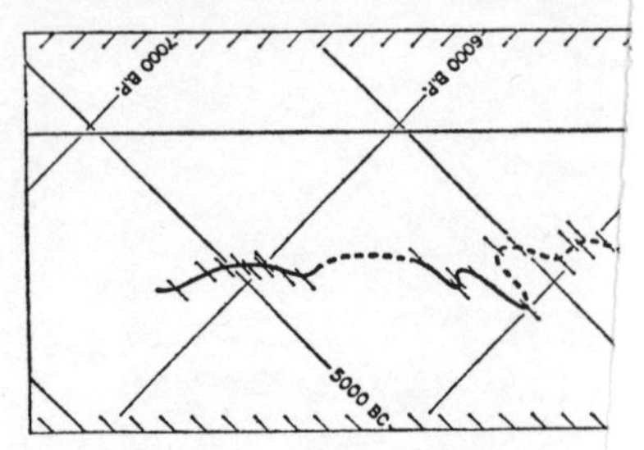

Bristlecone pine dates in calendar ye

Conventional radiocarbon dates in ra

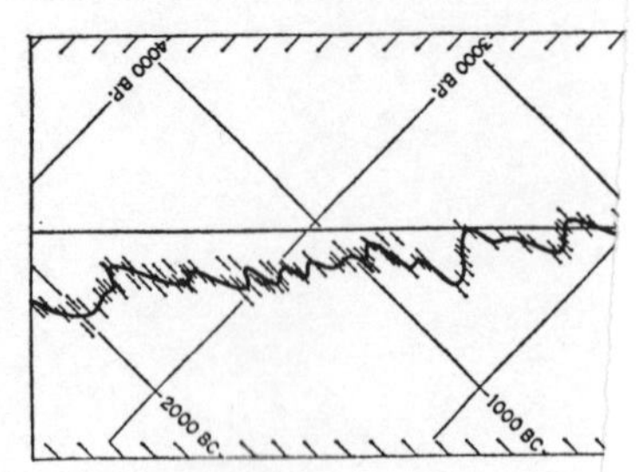

Bristlecone pine dates in calendar

Bibliographies

1 Science and Archaeology

BIBBY, GEOFFREY, *The Testimony of the Spade*, Collins, 1956; Knopf, New York, 1956.

CLARK, GRAHAME, and PIGGOTT, STUART, *Prehistoric Societies*, Hutchinson, 1965.

DANIEL, GLYN, *A Hundred Years of Archaeology*, London, 1950.

—, *The First Civilizations*, Penguin Books, 1971.

HAWKES, JACQUETTA, *The World of the Past*, Thames & Hudson, 1963; Knopf, New York, 1963.

ROE, DEREK, *Prehistory*, Macmillan, 1970

STERN, PHILIP VAN DOREN, *Prehistoric Europe*, George Allen & Unwin, 1970.

WOOLLEY, SIR LEONARD, *Digging up the Past*, Penguin Books, 1960.

2 The Importance of Dating

BIBBY, GEOFFREY, *The Testimony of the Spade*, Collins, 1956.

—, *Looking for Dilmun*, Collins, 1970.

BRYANT, VAUGHN M., and HOLZ, ROBERT K., 'The Role of Pollen in the Reconstruction of Past Environments', *Pennsylvania Geographer*, vol. 6, no. 1, February 1968, pp. 11–19.

COON, CARLETON, *Seven Caves*, Knopf, New York, 1957.

DANIEL, GLYN, *A Hundred Years of Archaeology*, London, 1950.

—, *The Megalith Builders of Western Europe*, Hutchinson, 1958.

GLOB, P. V., *The Bog People*, Paladin Books, Granada Publishing, 1971.

STERN, PHILIP VAN DOREN, *Prehistoric Europe*, George Allen & Unwin, 1970.

WOOLLEY, SIR LEONARD, *Digging up the Past*, Penguin Books, 1970.

ZEUNER, FREDERICK E., *Dating the Past*, 4th edition, Methuen, 1958.

4 The Clock Discovered

BANNISTER, BRYANT, 'Dendrochronology', in *Science in Archaeology*, D. Brothwell and E. Higgs, eds.,Thames & Hudson, 1963.

'Andrew Ellicott Douglass', *The Year Book of the American Philosophical Society*, 1965, pp. 121–5.

FRITTS, HAROLD C., 'Growth Rings of Trees, Their Correlation with Climate', *Science*, vol. 154, no. 3752, 25 November 1966, pp. 973–9.

LIBBY, W. F., *Nobel Prize Address* (1960), Stockholm, 1961.
—, 'Radiocarbon Dating', *Endeavour*, vol. 13, no. 49, January 1954, pp. 5–16.
—, 'Radiocarbon Dating', *Philosophical Transactions of the Royal Society*, A.269, 1970, pp. 1–10.
—, 'Radiocarbon Dating', *Science*, vol. 133, 3 March 1961, pp. 621–9.
LIBBY, W. F., ANDERSON, E. C., and ARNOLD, J. R., 'Age Determination by Radiocarbon Content: World-wide Assay of Natural Radiocarbon', *Science*, vol. 109, no. 2827, 4 March 1949, pp. 227–8.
LIBBY, W. F., and ARNOLD, J. R., 'Age Determination by Radiocarbon Content: Checks with Samples of Known Age', *Science*, vol. 110, no. 2869, 23 December 1949, pp. 678–80.
ZEUNER, FREDERICK E., *Dating the Past*, 4th edition, Methuen, 1958.

5 The Clock Reset

FERGUSON, C. W., 'Bristlecone Pine: Science and Esthetics', *Science*, vol. 159, no. 3817, 23 February 1968, pp. 839–46.
GREY, DONALD C., in an article in *Journal of Geophysical Research*, vol. 74, no. 26, 1 December 1969.
OLSSON, INGRID U., ed., *Radiocarbon Variations and Absolute Chronology*, Almqvist & Wiksell, Stockholm, and Wiley Interscience, New York, 1970. (The Proceedings of the Twelfth Nobel Symposium.)
SWITSUR, V. R., 'The Radiocarbon Calendar Recalibrated', *Antiquity*, vol. 47, no. 186, June 1973, p. 132.

7 More Clocks: The Breakthrough

AITKEN, M. J., 'Dating by Archaeomagnetic and Thermoluminescent Methods', *Philosophical Transactions of the Royal Society*, London, A.269, 1970, pp. 77–80.
AITKEN, M. J., ZIMMERMAN, D. W., FLEMING, S. J., and HUXTABLE, J., 'Thermoluminescent Dating of Pottery', in Ingrid U. Olsson, ed., *Radiocarbon Variations and Absolute Chronology*, Almqvist & Wiksell, Stockholm, and Wiley Interscience, New York, 1970.
BROTHWELL, DON, and HIGGS, ERIC, eds., *Science in Archaeology*, 2nd edition, Thames & Hudson, 1969.
FLEISCHER, ROBERT L., PRICE, P. BURFORD, and WALKER, ROBERT M., in Brothwell and Higgs, op cit., Section 3.
FRIEDMAN, IRVING, SMITH, ROBERT L., and CLARK, DONOVAN, in Brothwell and Higgs, op. cit.
GENTNER, W., and LIPPOLT, H. J., in Brothwell and Higgs, op. cit., Section 6.
HALL, E. T. in Brothwell and Higgs, op. cit.
HAWKES, J. and C., *Prehistoric Britain*, Chatto & Windus, 1947.

JOHNSEN, S. J., DANSGAARD, W., CLAUSEN, H. B., and LANGWAY, C. C., 'Oxygen Isotope Profiles through the Antarctic and Greenland Ice Sheets', *Nature*, vol. 235, 25 February 1972, pp. 429–34.

—, 'Climatic Oscillations 1200–2000 A.D.', *Nature*, vol. 227, 1 August 1970, pp. 482–3.

JOHNSON, FREDERICK, 'Radiocarbon Dating and Archaeology in North America', *Science*, vol. 155, 13 January 1967, pp. 165–9. (This is a revision of his paper 'The Impact of Radiocarbon on Archaeology' at the Sixth International Conference on Radiocarbon and Tritium Dating, Washington State University, June 1965.)

OAKLEY, K. P., 'Analytical Methods of Dating Bones', Brothwell and Higgs, op. cit., Section 1.

8 Revolution in a Revolution

BINFORD, L. R., and BINFORD, S. R., eds., *New Perspectives in Archeology*, Aldine, Chicago, 1968: especially the chapter 'Post Pleistocene Adaptations'.

CLARK, HELENA H., 'The Origin and Early History of the Cultivated Barleys', *The Agricultural History Review*, vol. 15, part 1, 1967.

CLUTTON BROCK, JULIET, in *The Domestication and Exploitation of Plants and Animals*, P. Ucko and G. Dimbleby, eds., Duckworth, 1969.

FLANNERY, KENT V., 'Origins and Ecological Effects of Early Domestication in Iran and the Near East', in *The Domestication and Exploitation of Plants and Animals*, P. Ucko and G. Dimbleby, eds., Duckworth, 1969, pp. 73–101.

GRIGSON, CAROLINE, in *The Domestication and Exploitation of Plants and Animals*, P. Ucko and G. Dimbleby, eds., Duckworth, 1969.

HARLAN, J. R., in *Archaeology*, vol. 20, no. 3, 1967.

HAWKES, J. G., 'The Ecological Background of Plant Domestication', in *The Domestication and Exploitation of Plants and Animals*, P. Ucko and G. Dimbleby, eds., Duckworth, 1969, pp. 17–31.

HELBAEK, H., Appendix I 'Plant Collecting, Dry Farming and Agriculture in Prehistoric Deh Luran', in 'Prehistory and Human Ecology of the Deh Luran Plain', by F. Hole, K. V. Flannery and J. A. Neely, *Memoirs*, University of Michigan, Museum of Anthropology.

HELBAEK, H., 'Palaeo-Ethnobotany', in *Science in Archaeology*, D. Brothwell and E. Higgs, eds., Thames & Hudson, 1969, Section 18.

RILEY, RALPH, in *The Domestication and Exploitation of Plants and Animals*, P. Ucko and G. Dimbleby, eds., Duckworth, 1969.

—, 'Wheat-breeding and the Behaviour of Chromosomes', *New Scientist*, vol. 17, pp. 698–700.

UCKO, PETER J., and DIMBLEBY, G. W., eds., *The Domestication and Exploitation of Plants and Animals*, Duckworth, 1969.

ZOHARY, DANIEL, 'The Progenitors of Wheat and Barley in Relation to Domestication and Agricultural Dispersal in the Old World', in Ucko and Dimbleby, op. cit., pp. 35–47.

9 Man and Maize

BYERS, DOUGLAS S., ed., *The Prehistory of the Tehuacan Valley*, University of Texas Press (for the Robert S. Peabody Foundation), 1967: especially the first and last chapters of the first volume, which are written by R. S. MacNeish.

KAPLAN, L., a letter in *Science*, vol. 179, 5 January 1973, p. 77.

MACNEISH, R. S., 'Ancient Mesoamerican Civilisation', *Science*, vol. 143, no. 3606, 7 February 1964, pp. 531–7.

—, 'Early Man in the Andes', *Scientific American*, vol. 224, no. 4, April 1971, pp. 36–46.

MANGELSDORF, PAUL C., MACNEISH, R. S., and GALINAT, WALTON C., 'The Domestication of Corn', *Science*, vol. 143, no. 3606, 7 February 1964, pp. 538–45.

WILKES, H. GARRISON, 'Maize and its Wild Relatives', *Science*, vol. 177, no. 4054, 22 September 1972, pp. 1071–7.

10 Traces, and How to Find Them

AITKEN, MARTIN, 'Magnetic Location', in *Science in Archaeology*, edited by Brothwell and Higgs.

—, 'The Proton Magnetometer', Proceedings of Convegno internazionale sulla technica e il diritte nei problemi della odierna archeologica', Rome, 1962.

ANDERSON, DOUGLAS A., 'A Stone Age Campsite at the Gateway to America', *Scientific American*, Vol. 218, No. 6, June 1968, pp. 24–9.

ASPINALL, A., WARREN, S. E., CRUMMET, J. G., and NEWTON, R. G., 'Neutron Activation Analysis of Faience Beads', *Archaeology*, vol. 14, part 1, February 1972, pp. 27–41.

BROTHWELL, DON, and HIGGS, ERIC, *Science in Archaeology*, 2nd edition, Thames & Hudson, 1969.

CANN, J. R., DIXON, J. E., RENFREW, C., 'Obsidian Analysis and the Obsidian Trade', in Brothwell and Higgs, op. cit., Section 51.

—, 'Obsidian and the Origins of Trade', *Scientific American*, vol. 218, no. 3, March 1968, pp. 38–46.

CATLING, H. W., and MILLETT, A., 'A Study of the Inscribed Stirrup Jars from Thebes', *Archaeometry*, vol. 8, 1965, pp. 3–85.

CLARK, ANTHONY, 'Resistivity Surveying', in Brothwell and Higgs, op. cit.

GUMERMAN, GEORGE J., and LYONS, THOMAS R., 'Archaeological Methodology and Remote Sensing', *Science*, vol. 172, 9 April 1971, pp. 126–31.

HAWKES, J. and C., *Prehistoric Britain*, Chatto & Windus, 1947.

Masca Newsletter, University of Pennsylvania, Museum Applied Science Center, 1966–71.

PATTON, and MILLAR, *Science*, vol. 169, 1970, p. 760.
SHOTTON, F. W., 'Petrological Examination', in Brothwell and Higgs, op. cit., Section 50.
Times, The, 15 August 1973, for results from Franchthi cave.
WOOLLEY, SIR LEONARD, *Digging up the Past*, Penguin Books, 1970.

11 Computers and the Past

HAWKINS, G. S., *Stonehenge Decoded*, Souvenir Press, 1966.
KATZ, FRIEDRICH, *The Ancient American Civilizations*, Weidenfeld & Nicolson, 1972.
MARSHACK, ALEXANDER, 'Upper Palaeolithic Notation and Symbol', *Science*, vol. 178, no. 4063, 24 November 1972, pp. 817–27.
Masca Newsletter, University of Pennsylvania, Museum Applied Science Center, December 1971.
RENFREW, A. C., ed., *The Explanation of Culture Change: Models in Pre-History*, Duckworth, 1974.
THOM, A., *Megalithic Sites in Britain*, Oxford University Press, 1967.
—, *Megalithic Lunar Observatories*, Oxford University Press, 1971.
—, an article in *Journal for the History of Astronomy*, vol. 2, part 3, October 1971.
—, an article in *Journal for the History of Astronomy*, vol. 3, part 1, February 1972.

12 The Remains of Man

BIDDLE, MARTIN, *Health in Medieval Winchester*, University of Exeter, 1966.
BROTHWELL, DON, 'Community Health as a Factor in Urban Cultural Evolution', in *Man, Settlement and Urbanism*, Proceedings of Symposium at the Institute of Archaeology, edited by Ucko, Tringham and Dimbleby, Duckworth & Co., 1972.
BROTHWELL, DON, and HIGGS, ERIC, eds. *Science in Archaeology*, 2nd edition, Thames & Hudson, 1969: especially the editors' introductory chapter 'Scientific Studies in Archaeology'.
BROTHWELL, DON, MORSE, DAN, and UCKO, PETER J., 'Tuberculosis in Ancient Egypt', *American Review of Respiratory Diseases*, vol. 90, no. 4, October 1964.
BROTHWELL, DON, and BROTHWELL, PATRICIA, *Food in Antiquity*, Thames & Hudson, 1969: especially the chapter 'Diet and Disease'.
COCKBURN, T. A., in *Science*, vol. 133, 1961.
COON, CARLETON, *The Origin of Races*, Jonathan Cape, 1963; Knopf, New York, 1962, p. ix.
GENOVES, SANTIAGO, 'Sex Determination in Earlier Man', and 'Estimation of Age and Mortality', in Brothwell and Higgs, op. cit., Sections 37 and 38, pp. 429–53.

GOLDSTEIN, MARCUS S., 'The Palaeopathology of Human Skeletal Remains', in Brothwell and Higgs, op. cit., Section 41, pp. 480–90.

Newsweek, 13 November 1972.

WELLS, L. H., 'Stature in Earlier Races of Mankind', in Brothwell and Higgs, op. cit., Section 39, pp. 453–68.

ZEUNER, F., *Dating the Past*, 4th edition, Methuen, 1958.

13 Fakes

AITKEN, M., UCKO, P., SCHWEIZER, in *Archaeometry*, vol. 13, part 1, 1971, pp. 89–141.

JUCKER, H., FLEMING, S., RIEDERER, J., in *Archaeometry*, vol. 13, part 1, 1971, pp. 143–67.

OAKLEY, KENNETH P., 'Analytical Methods of Dating Bones', in Brothwell and Higgs, op. cit., pp. 35–45.

OAKLEY, KENNETH P., and WEINER, J. S., 'Piltdown Man', *American Scientist*, vol. 43, 1955; Bobbs-Merrill, New Haven, Conn.

PEACOCK, D. P. S., in *Antiquity*, vol. 47, no. 186, June 1973, pp. 138–40.

14 Crisis in the Profession

LONGACRE, WILLIAM A., 'Current Thinking in American Archeology', *Bulletins of the American Anthropological Association*, vol. 3, part 3, no. 2, 1970.

—, A chapter in *New Perspectives in Archeology*, L. R. Binford and S. E. Binford, eds., Aldine, Chicago, 1968.

15 Crisis in the Theory

BERGER, RAINER, and PROTSCH, REINER, 'Earliest Radiocarbon Dates for Domesticated Animals', *Science*, vol. 179, no. 4070, 19 January 1973.

BINFORD, L. R., and BINFORD, S. R., eds., *New Perspectives in Archeology*, Aldine, Chicago, 1968.

BROTHWELL, DON, and HIGGS, ERIC, eds., *Science in Archaeology*, 1st edition, Thames & Hudson, 1963.

CHILDE, V. GORDON, in *Antiquity*, vol. 32, no. 70, 1958.

PIGGOTT, STUART, in *The Domestication and Exploitation of Plants and Animals*, ed. Ucko, P. and Dimbleby, G., Duckworth, 1969, p. 559.

RENFREW, COLIN, *Before Civilization*, Jonathan Cape, 1973; Knopf, New York, 1973.

—, 'Carbon 14 and the Prehistory of Europe', *Scientific American*, vol. 225, no. 4, October 1971, pp. 63–72.

—, 'The Revolution in Prehistory', *Listener*, vol. 85, no. 2180, 7 January 1971.

16 The New Archaeology

FRENCH, DAVID, *Report of the English Archaeological Institute in Ankara*, 1971.

HILL, JAMES, 'Broken K Pueblo', in *New Perspectives in Archeology*, L. R. Binford and S. R. Binford, eds., Aldine, Chicago, 1968.

RENFREW, COLIN, *Before Civilization*, Jonathan Cape, 1973; Knopf, New York, 1973.

RENFREW, A. C. ed., *The Explanation of Culture Change: Models in Pre-History*, Duckworth, 1974.

17 Archaeology and History

BIDDLE, MARTIN, 'Winchester, the Archaeology of a City', *Science Journal*, March 1965, pp. 55–61.

—, 'Winchester 1961–68', Chateau Gaillard IV, Conference, Ghent, 18–25 August 1968.

—, *Archaeology and the History of British Towns*, Sartryck ur Res Mediaevales, no date.

TUCK, JAMES A., in *Scientific American*, vol. 224, no. 2, February 1971, pp. 32–41.

Index

A Note About the Author

David Wilson was born in 1927 in Rugby, Warwickshire, England. Educated at Ampleforth College, York, and Pembroke College, Cambridge, Mr. Wilson studied mathematics, physics, and history, receiving a B.A. and an M.A. from Cambridge University. A newspaper reporter from 1950 to 1955, he joined the BBC news department in 1956, and has been their Science Correspondent since 1963. In addition to broadcasting regularly on BBC radio and television, Mr. Wilson has published a number of articles in *The Listener* and is the author of *Body and Antibody* and a forthcoming history of the discovery of penicillin. He is married, and the father of four children.

A Note on the Type

The text of this book was set in Plantin, a typeface cut in 1913 by the Monotype Corporation Limited. Though the face bears the name of the great Christopher Plantin, who in the latter part of the sixteenth century owned, in Antwerp, Europe's largest printing and publishing firm, it is a rather free adaptation of designs made for that firm by Claude Garamond. With its strong, simple lines, Plantin is a no-nonsense face of exceptional legibility.

Binding design by Gwen Townsend